Voyage
TO THE
Whales

Voyage
TO THE
Whales

HAL WHITEHEAD

CHELSEA GREEN PUBLISHING COMPANY
POST MILLS, VERMONT

In memory of
Virginia Paine
and Ada Redshaw

Library of Congress Cataloging-in-Publication Data

Whitehead, Hal.
    Voyage to the whales.

    Bibliography: p.
    1. Sperm whale—Indian Ocean. 2. Whales—Indian Ocean
    3. Scientific expeditions—Indian Ocean.
    I. Title.

QL737.C435W47 1990   599.5'3   89-17330
ISBN 0-930031-25-3

Published by arrangement with Stoddart Publishing Co. Limited,
Toronto, Ontario, Canada

CONTENTS

PREFACE

TULIP'S VOYAGE TO THE SPERM WHALES of the Indian Ocean was an immensely powerful experience for all of us who sailed on board her. It introduced me to one of the largest and most unusual animals on earth in more intimate terms than I had ever known any other. We sailed with the sperm whales night and day; we swam with them and were present when one was born. But the whales, although the goal of the *Tulip* Project, were only one part of the voyage. We encountered other unusual and intriguing animals, sailed to strange and exciting countries and fought our way through an almost endless sequence of obstacles. During the voyage close relationships grew between us, and were sometimes destroyed. For me it was both a wonderful but immensely difficult experience.

As I started to write this book, partly as a catharsis, I wanted to describe "what it was really like out there," from the most mundane routine of living on a small boat at sea, to the finest and foulest moments. As I showed people drafts, it became clear that the detailed descriptions of bureaucratic complexities, of engine failure and personal relationships, were usually boring for those who had not taken part in the project and sometimes offensive for those who had. So this book focuses more on the exceptional and exciting parts of our voyage, on the whales themselves, the ocean and our more dramatic disasters. Most readers will have themselves experienced, in some form or other, the frustrations of officialism and persistently unreliable technology, as well as the intensity of personal relationships when people are living and working in very close quarters. If you can magnify these experiences and lay them over what I describe of our voyage, I hope that you will gain a feeling for both the wonder and difficulty of our times with the whales.

This book is not intended to be a scientific description of our research or its results, or a treatise on whale biology. Readers who are interested can find more complete and formal information about both sperm whales and whales in general in the books and papers listed in the bibliography. The bibliography also contains a list of many of the publications, both scientific and "popular," that describe the *Tulip* voyage and the research on living sperm whales that succeeded it.

I have taken a few liberties with chronology, moving some events forward or backward by a few days in order to maintain a narrative flow, but everything described herein did happen. I have also used photographs taken during the sperm whale research that succeeded the *Tulip* study to illustrate some aspects of sperm whale behavior for which no good photographs from *Tulip* exist.

The *Tulip* voyage, and this book, would not have taken place without inspiration, help and simple kindnesses from a large number of people. I thank them all, but especially the permanent crew of *Tulip*: Jonathan Gordon, Gay Alling, Nicola Rotton, Margo Rice, Martha Smythe, Chris Converse, Phil Gilligan, Caroline Smythe, Linda Weilgart, Nicola Davies and Vassili Papastavrou. The late Pieter Lagendijk, Khamis al-Farsi, Lex Hiby, Roger Payne, Cedric Martenstyn and Flip Nicklin joined us on *Tulip* for different periods, each adding new and valuable directions to our work and shared lives. The people of the Netherlands generously funded the research through the World Wildlife Fund and the International Union for the Conservation of Nature and Natural Resources. Cetacean Society International supported us both morally and financially. Sidney Holt largely set the study up and together with Sidney Brown, John Harwood, Lex Hiby, Stephen Leatherwood, James

Mead, Patricia Moehlman, Captain W.F.J. Mörzer Bruyns, Roger Payne, William Perrin, Graham Ross, Michael Tillman, Peter Tyack, William Watkins and Lyall Watson provided scientific advice. My parents, Denis and Frances Whitehead, supported us throughout, and most generously let us use their *Elendil-Tulip* for three long years, by the end of which she was unfortunately quite a different boat. Jon Lien, Roger Payne and the Sea Mammals Research Unit kindly loaned equipment. Robert Dalpech helped us with diplomatic arrangements. In Greece Martin Jenkins helped us outfit *Tulip*, and in Djibouti Yves Graille sewed our jib back together. Among the many Omanis who gave us hospitality and assistance were Mohamed al-Barwani, Mr. Ali, Khamis al-Farsi, the director general of Fisheries, the captain and crew of *Shebab Oman* and the staff of the Omani National Fish Corporation. In Sri Lanka we were very fortunate to have the friendship of Brian Lourenz and the assistance of his company, Constellation Yachts. Many other Sri Lankans and residents of Sri Lanka showed us hospitality or helped us in different ways. They include the personnel at the late Banno Lanka, Ltd., Tissa Amaratunga, Arthur C. Clark, Chandima de Alwis, Ranjen Fernando of the Wildlife and Nature Protection Society of Ceylon, Rodney Jonklass, Cedric Martenstyn, Mike and Regina Santerre, Mike Smith, Rosemarie Sommers and the Tainter family. The National Aquatic Resources Agency gave us permission to work in Sri Lanka and helped us in many ways, and the Department of Wildlife Conservation, especially Lynn de Alwis and Mr. Packeer, were very helpful in providing permits so that Gay's dolphin skulls could be exported. W.P. Thunga Prema, E.R. Tranchell, M. Cyril Fernando, L.W. Arthur and S. Kugarajah, who collected data within their communities, were instrumental in the success of Gay's dolphin by-catch study. The governments of the Maldives and the Seychelles kindly gave us permission to work in their waters. James Bird, Eleanor Dorsey, Linda Guinee, Patricia Harcourt, Kate O'Connell and Victoria Rowntree helped us analyze the data collected during the study.

Elizabeth Kemf of World Wildlife Fund and William Graves of the National Geographic Society encouraged me to write about the *Tulip* Project. This book has been greatly improved by the detailed reviews of several drafts by Barbara Blouin and Linda Weilgart. I also received most helpful comments on sections of the book from Carol Burnes, Gay Alling, Jonathan Gordon and other members of the permanent crew of *Tulip*. The encouragement and advice of Larry Hoffman, my agent, have been vital in allowing this book to reach publication. Illustrative photographs have kindly been provided by Gay Alling, Bruce Coleman, Ltd., Chris Converse, Jonathan Gordon, Pieter Lagendijk (through World Wildlife Fund Netherlands), the National Geographic Society, Flip Nicklin, Margo Rice, Linda Weilgart and World Wildlife Fund International.

Above all, I am grateful to Linda Weilgart for her loving support through the difficulties and separation of the project itself and then the tedious writing and rewriting of this book.

All royalties due to the author from sale of this book will be used for further whale research.

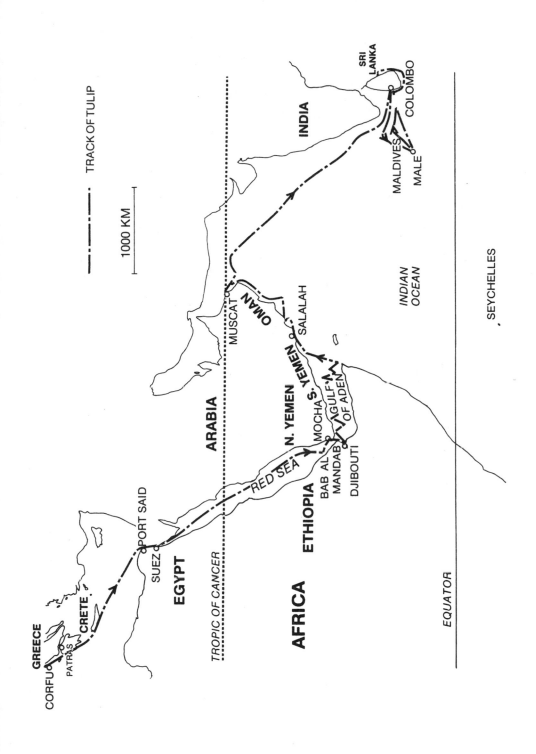

PROLOGUE
Off Colombo, Sri Lanka
November 12, 1983

IN THE LATE AFTERNOON WE CLOSE ON a single sperm whale. The big battleship-gray animal lies at the surface, motionless, the waves lapping the thin strip of back that is all we see above the surface. The whale appears like a straight log with two smooth bulges rising from either end. A weak misty spray appears from one of the bulges, the blowhole. There is no reaction as I steer *Tulip* alongside so that Lindy can photograph the dorsal fin — the other bulge. The fin is topped with a callus, sign of a female. The photographs taken, we hang back and wait for her to dive. We wait and wait. The whale just lies at the surface, blowing regularly every fifteen seconds. After ten minutes Flip, the photographer from *National Geographic*, asks, "Would it be okay if I got in the water? The light isn't too great, but I might be able to shoot a silhouette."

He checks his camera, slips into the water and swims over to the whale, who seems totally unperturbed. He swims around her, taking photographs. The whale shows no sign of movement. Then Flip lifts his flippers into the air, diving for a silhouette shot. The whale finally responds: she raises her flukes (tail) and dives after Flip. Soon they are both back at the surface, and Flip quickly uses up his roll of film. He touches the whale. There is no reaction. He then rubs the length of her body, with no reaction, and returns to the boat.

"That was really weird," Flip reports as he changes his film. "I swam right up to the eye and saw neither aggression nor interest. I've never known a whale like this one."

"Have you ever had a whale follow you down on a dive?" Phil asks.

"No, I usually follow them!"

1

"The whale seems sick," I suggest.

"She's probably pregnant and there are complications with the birth. Was it definitely a female?" Lindy asks Flip, who has seen the genital area.

"I think so. She did seem fat."

"She certainly looked wide from here on deck," Lindy adds. "Poor baby, she seems so totally listless. She's probably in terrible pain."

Flip returns to the water, this time accompanied by Phil. They swim over to the whale, who is now moving very slowly on a straight course. As the light fades, they photograph each other beside the strangely passive animal. She remains almost totally undemonstrative, although other sperm whales glide by, twisting their flukes as they maneuver past the divers.

The whale seems to care so little about anything external that we are able to sail *Tulip* right beside her. From six feet away we watch the dark smooth forehead, the hugely wrinkled back mottled with peeling skin, which is being shed in large sheets. The powerful blowhole muscles hunch up to open and allow her breath to escape. The mist from the blow wets our cameras, notebooks and clothes, and the oily-sea smell of the inside of a whale hangs over *Tulip*'s cockpit.

Finally the whale turns, lifting the swimmers' safety rope over her forehead. Phil, who had been clinging to the end of the rope, drops off, while Flip, who is closer to *Tulip*, rises from the water on the whale's back, before tumbling over the other side. Now she is slowly heading directly toward us in *Tulip*'s cockpit; she is only ten feet away. Surely she will turn; they always do. But, no, the whale blunders right into *Tulip*, just beneath us. There is a thud; our little home shudders. Surely, surely she will turn now. But she keeps right on swimming. *Tulip* rises slightly from the water and heels over as the whale passes beneath; we can hear the huge animal scraping on the hull. She emerges from the other side, still swimming slowly and steadily.

A sorrowful concern is in the air as Flip and Phil clamber

aboard. We hang back and watch the whale plodding slowly toward the sunset.

"We've *got* to stay with the group through the night to make sure she's okay and has her baby safely," says Lindy.

Through the evening we track the whales moving slowly southwest with our directional hydrophone. But the wind is rising; at midnight it is approaching gale force, and whale tracking is becoming almost impossible. I give the order to sail back to Colombo. At one in the morning Lindy writes in the log: "Saw three shooting stars and a bloodred moonset. Hope that means poor sweet mama whale is okay, and her *baby*!"

Part I

From the Mediterranean through the Red Sea to the Indian Ocean
Winter 1981 - 1982

1

SOUTH OF CRETE

November 15 - December 1

IT IS A GALE — A SPECTACULAR GALE with the sun sparkling off the deep swells. A wave crashes against *Tulip*'s topside, one inch behind me. She lurches, then steadies, raises her bow and accelerates. She shudders slightly as the power of the wind lifts her hull to surf over the seas in a storming, exhilarating charge. Inwardly I whoop with joy. "Off to the Indian Ocean!"

During the past weeks we feel as though we have been delayed and frustrated at every step, but, at last, this wild Cretan gale is saying, "If you can ride me, I will take you there."

The four members of *Tulip*'s crew are scattered about her hull: Jonathan huddled in the cockpit, Nic stretched out in the forepeak, while Pieter and I occupy opposite corners of the main cabin. Our bodies are contorted in attempts to minimize the debilitating discomfort of *Tulip*'s twisting passage. There is no conversation; each is alone with the gale.

Pieter Lagendijk, the big blond Dutchman, fills the starboard bunk, his feet buttressed against the midships bulkhead, an open book facedown on his chest. He stares up through a spray-spattered skylight at the mast carving drunkenly through the bright Grecian sky. Pieter is the representative of World Wildlife Fund Netherlands, the organization that has made this voyage and study possible. We are sailing for the Indian Ocean to study the living sperm whale.

While the major impetus behind the project has been Dutch, the substance of our voyage has flowed from a number of unlikely sources in several lands. In 1979 the government of the Seychelles, a small country in the central Indian Ocean, rallied the other nations in the region and proposed to the International Whaling Commission (I.W.C.) that the Indian Ocean be declared a sanctuary for whales.

When the concept of a sanctuary was presented by the Seychelles to the I.W.C., there were several opposing undercurrents, as well as openly expressed "causes for concern." The major powers were alarmed at any alliance among Indian Ocean nations, even to protect whales. For a number of years the countries of the region had complained about the military presence of distant powers (particularly the American base on the British-controlled island of Diego Garcia) in what the Indian Ocean states wish to be "a sea of peace." Any marine alliance of these smaller nations might harm American/-British/French/Soviet "strategic interests."

More obviously, there was opposition from the Japanese and Russian whalers, who had a substantial fishery for small minke whales in the southern part of the Indian Ocean, off the Antarctic continent. These countries vigorously protested the creation of a sanctuary that would halt their lucrative slaughter.

The sanctuary was also opposed on scientific grounds: most of the older whale scientists had been brought up on the bloody flensing decks of factory ships and whaling stations. They had waded through gory carcasses, taking samples and making measurements, clues to the natural history of the whale. Many of these scientists believe that if no whales are killed, nothing will be learned of them; whale science, they contend, could not exist without a whaling industry and carcasses to mark, count and measure. However, this familiar notion has been largely discounted by the emergence of a new kind of whale science, which has come to be called "benign research." Over the past twenty years, men and women have been working from cliffs, small boats, or light airplanes, usually off the American continents, trying to scratch at the mysteries of living whales. While there have been many wasted days, frustrating times and disillusioned participants, the results wrung from some parts of this diffuse effort have been remarkably detailed. Particularly illuminating studies have been done on the right whales off Argentina, gray whales off the west coast of North America, orcas (or killer whales) near Vancouver Island and Washington state and humpback whales on both sides of the North American continent. In 1979, however, there was still some debate among the I.W.C. scientists about the validity of benign research and how applicable it was to other species of whale.

After rounds of bargaining the I.W.C. came to a typically political compromise: most of the Indian Ocean would be made a sanctuary, but the portion south of latitude fifty-five degrees south would be excluded. This, of course, was where almost all whaling was taking place. Requesting a rider, the I.W.C. called for research on the protected whales; it would have to be benign research.

World Wildlife Fund Netherlands was quick to respond. In November 1980, at the televised opening of a flower auction hall in Rijnsburg, Prince Bernhard of the Netherlands made an appeal on behalf of World Wildlife Fund for money to carry out a research project on the protected whales of the Indian Ocean. The Dutch people, who have a reputation for environmental awareness, responded generously. The proceeds from that one night were sufficient to launch a major research project.

During the winter of 1980-1981, I was in the throes of completing my Ph.D. thesis ("The Behaviour and Ecology of the Humpback Whale in the Northwest Atlantic") at Cambridge University. Also at Cambridge was Dr. Sidney Holt, a particularly distinguished champion of the whales. During the 1960s he had been a member of the Committee of Three, experts on the population dynamics of fish who were called in by the I.W.C. when it became clear that the whale stocks were in grave danger. Since then he has been an eloquent advocate of marine conservation, especially of marine mammals. He has also been influential in guiding the scientific and political actions of the conservation community, and had a major role in setting up the Indian Ocean whale sanctuary. There could be few better advisers for a whale scientist trying to arrange the next part of his life.

"Have you ever considered studying sperm whales in the Indian Ocean?" he asked me, talking slowly.

"No," I said with a gulp.

Most of the little that I knew of sperm whales came from having read *Moby-Dick* some years before. In the course of several years of whale research at sea, I had only sighted sperm whales twice. And to me the Indian Ocean was a mysterious region of which I had only a superficial geographic conception.

"The Dutch World Wildlife Fund have raised some money for research in the Indian Ocean sanctuary," Sidney explained, "and I think it could be used to study sperm whales. You might want to consider it."

"Why me?" was my first reaction. Nobody knew where to go in the Indian Ocean to study living sperm whales. Furthermore, they are creatures of deep water, rarely coming close to land. Unlike most of the scientists involved in benign whale research, who watch whales from cliffs, small airplanes, or outboard-motor-powered boats, I usually work from small oceangoing sailing yachts. This allows us to study whales hundreds of miles from land and to survey different possible research areas over a whole ocean.

As I returned to the frustrations of my thesis, the thought of research on Indian Ocean sperm whales hung in the back of my mind, but it was usually overshadowed by the immediate struggle of analyzing data and presenting my results, as well as by the prospect of a summer sailing with the humpback whales off West Greenland.

Some weeks later Sidney Holt invited me to a meeting in London at which possible uses of the W.W.F. money were to be discussed. At the meeting were Sidney, two representatives of World Wildlife Fund Netherlands, one from World Wildlife Fund International, three famous whale scientists and me. One of the three scientists was Dr. Roger Payne of the New York Zoological Society, who has become virtually the "father" of benign whale research. It was he who discovered the intricate song of the humpback whale, and his remarkably detailed long-term study of the right whales off Peninsula Valdes, Argentina, has been the model for the new whale science. I worked for him at his laboratory in Lincoln, Massachusetts, during a stimulating year and a half before returning to England for my graduate degrees. Roger has an unequaled ability to somehow know what whales are actually doing, and he has energetically combined hard science with the promotion of conservation. He has been the major inspiration for my generation of whale scientists.

Sidney Holt introduced the purpose of the meeting: to find the most appropriate use of the Dutch funds. We were asked to discuss various projects. Since my proposal was the simplest, I was asked to start. I explained how I would begin to study the behavior of the sperm whale in the Indian Ocean.

"I would use a sailing boat thirty-two to forty-two feet long with a good engine, with a crew of four to five people who were both scientists and sailors. I would divide the study into three field seasons of about four months each, spread over three years so that we could choose the times of best weather and get a chance to analyze our data between seasons. During the first season I would survey the northwest Indian Ocean, trying to find the most suitable research area, and investigate different techniques of studying sperm whales. If we find sperm whales and a means to study them, then during the second and third seasons we will concentrate on the whales within our study area, develop our methods further and start to build up a data base of information on living sperm whales. We will also try to study other species of whale and dolphins that we encounter."

The other scientists considered the plan.

"Well, Hal, I think your proposal is scientifically important, but the World Wildlife Fund needs publicity," said one. "A standard fiberglass yacht sailing along with a few sperm whale backs in the distance won't make a very good TV film."

I agreed; my kind of research is not particularly photogenic.

"I think we should dive with the whales," said another. "It's dramatic, and there's a wealth of information to discover."

"The link with Holland is important. I think we should use a traditional Dutch sailing barge," added the first scientist.

"With the crew wearing Dutch national dress and clogs!" quipped Roger.

Having no experience of, and little inclination for, the publicity side of whale research, and being a sailor, not a dive I seemed to be ruled out.

However, the representatives of World Wildlife Fund Netherlands, who had stayed silent, suddenly picked up my proposal and announced, "We like Hal's project."

During the rest of the meeting we discussed how publicity value and "Dutch content" might be incorporated into the study. It was suggested that we might change the name of our research vessel to reflect the Dutch origins of the expedition.

"What about *Petal?*" suggested the helpful first scientist.

I cringed as Roger gave a sympathetic wink. Luckily the Dutch representatives did not seem overjoyed with *"Petal."*

"Well, perhaps *Tulip*, then?" the first scientist proposed.

The Dutch were more enthusiastic. "Yes. It symbolizes the flower auction hall and is almost the same word in Dutch as in English."

So now it is *Tulip* that will take us to the Indian Ocean.

I walked through the rainy London afternoon confused and unsure. There was the excitement of the unknown Indian

Ocean, of the fabled sperm whale, but "publicity value"? The size and potential complexity of the project were intimidating. And the heat — I love the cold, clear, arctic waters and generally feel oppressed in the tropics. Should not I remain with the familiar humpback and finback whales off the cliffs of Newfoundland?

On returning to Cambridge after the meeting at which the project was settled, I immediately telephoned Jonathan Gordon. I had met Jonathan a year earlier when he had been an undergraduate at Cambridge, studying zoology. A little above average height and slim with fair hair, a beard and inquisitive eyes, Jonathan looks the part of a field biologist. He is a sailor, diver and naturalist, whose consuming interest was once seals. Jonathan had traveled to the Azores islands one summer to see whether the endangered Mediterranean monk seals survived there. He quickly discovered they probably did not, and then set out to investigate the primitive sperm whale fishery the Azorians carried on. Despite the change of focus, the expedition was successful.

Jonathan wished to pursue his interest in marine mammals professionally, but there were no opportunities for graduate work on seals or whales in Britain, so he started a Ph.D. in Scotland, studying kelp-bed fish. After a frustrating year spent diving in the kelp for elusive fish, Jonathan was open to my suggestion to make a Ph.D. study of the Indian Ocean sperm whales. This meant, however, giving up his safe government grant and embarking on a project of extremely uncertain potential.

I was grateful when Jonathan called back and said he would come. His sailing ability would be important, and his experience with sperm whales would help compensate for my ignorance, but the principal benefit would be that he could carry much of the burden of planning the project and analyzing the data. I was so weary of my own dissertation that I did not wish to embark on another monumental analysis.

13

It is seven months later. Pieter, our "Dutch content," clutches the side of his bunk as *Tulip* lurches even more violently. The wind has risen and *Tulip* is a little overpowered; it is time to change gear. I struggle out of my niche and haul myself up into the cockpit. Jonathan, who is on watch, gives a wry grin and indicates the reefed mainsail with a jerk of the head. I nod. We clamber up to the mast. As Jonathan lowers the halyard, I bundle up the flogging canvas. With the sail safely stowed away we rest against the boom and watch the streaming foam. A shearwater wheels over the ocean at *Tulip*'s bow. Its wing tips teasingly brushing the waves, the neatly marked brown-and-white bird shows us how to live gracefully at sea.

Generally, the most uncomfortable position on *Tulip* is in the forepeak. In this stuffy triangle every movement of the boat is exaggerated; the topsides, which form its walls, heave on every surge, and a small part of every deck-sweeping wave dribbles through the hatch. In this claustrophobic, damp, flexing, roller coaster, Nic (Nicola) Rotton tries to rest. Tall and slim with dark hair, she is an experienced sailor, and was a member of Jonathan's Azores expedition. A steady, down-to-earth woman, Nic has had a varied work experience ranging from drama to teaching. She has already more than proved her worth; her common sense has been a stabilizing force during our recent trying times.

Nic is the first of several "permanent" crew members who will join Jonathan and me for one or more seasons. They will have to do much of the difficult, tedious work. Other "visiting" crew, like Pieter, will accompany us for a few weeks at different times. As we are now principally making a passage and will not dally long with any whales until we reach the Indian Ocean, we carry only four crew.

There is little talk; each is alone with his or her feelings — fondly remembering the stable land, exulting in the power of the ocean, or simply fighting seasickness.

Elendil-Tulip, which belongs to my parents, is a thirty-three-foot-long fiberglass oceangoing sloop. She was built in France to an English design, and although this cross-channel union may sound like a recipe for disaster, she is both well designed and well built. She is tough enough to have survived all that several North Atlantic gales have thrown at her, and is, for her size, fast and comfortable. For much of the next three years *Tulip* will be our home, the platform from which we will view the whales and the world.

I gaze out over the port rail. Twenty miles away rise the hazy snow-clad mountains of Crete, a reminder of simpler times when, nine years ago, I trekked those slopes alone, with no goal or timetable, wandering slowly through the Cretan mountains, sleeping in a cave, a monastery or a peasant hut. But now, swept eastward by this benevolent gale, we rush past Crete, already many days behind schedule; we *must* reach Djibouti by Christmas.

I had arrived on the Greek Island of Corfu in mid-November 1981. Jonathan, Nic and Pieter had been there two weeks and were working hard. *Tulip* had been hauled and painted. They had managed to find places to stow our enormous variety of gear: nets, a depth sounder, cameras, books, hydrophones, bottles, tape recorders, film, a video system with cameras, diving equipment, a satellite navigator, tools, radios, mountains of food, still and movie cameras, tapes, spare sails, batteries, a dinghy, and much more.

But there were still problems. Air freight had been mislaid, there were costly bureaucratic rituals to be performed over each item that did arrive, and the engine persistently overheated — a fault that four mechanics had been unable to cure. Soon after my arrival in Greece, I delved into the engine, and a few days later I experienced true joy in watching a steady temperature gauge.

We were about a week behind schedule when we sailed out of Corfu on November 23. I was happy to be back at sea, excited to be sailing for the Indian Ocean and the sperm whales. The land still inhabited me, its complexity and bureaucracy, but the simple strength of the sea was starting to soothe.

Just one day later, however, off the honey-colored island of Cephalonia, joy had turned to despair. The engine had ground to a halt. I fiddled with it, replaced parts and finally dismantled the ghastly green Jonah. There was water in the oil and in the cylinder, and a connecting rod had broken.

We pointed *Tulip* in the direction of Patras, the nearest large town, but for thirty hours she did not stir: we were becalmed. I tried to relax, to admire the fine islands that surrounded us, but our situation was too frustrating. We had traveled less than one hundred and twenty-five miles of the many thousand we hoped would take us to the whales. Our first season in the Indian Ocean, if we ever got that far, was slipping away.

The wind finally came; fitfully, unwillingly, the hazy air moved, and *Tulip*, too. We put up our new spinnaker, emblazoned with a giant panda, the logo of World Wildlife Fund (W.W.F.). Nic christened it "Wuff-Wuff." The name has stuck to the sail and, through association, to the organization.

Wuff-Wuff drew us to Patras, where we finally managed to arrange for a mechanic to look at the engine. A few hours later an old man arrived on a bicycle. He glanced at the engine, shook his head, muttered, "Bad . . . bad . . ." and pedaled off.

At this stage we had virtually no faith in being able to find a willing, competent mechanic. We considered the engine's dismal record, and the thousands of miles we were going to be traveling, into areas where there would probably be even less capable help available, and we decided to start afresh and buy a replacement.

I took the bus to Athens, visited a slick engine emporium and handed over a large sum of money. I was promised that a brand-new engine would be delivered immediately to Patras, where it would be installed by a "very competent mechanic." Two days later it arrived. Not all its peripheral parts looked completely unused, but we did not know how to check them, and were desperate to leave. The competent mechanic announced he was unable to install it for three days because of holidays and other pressing engagements. We haggled and persuaded. Finally he gave in, and he and his crew carried the engine down to the boat. With curses, yells and crowbars they forced the "new" engine in. I watched, furious, frustrated and sad. The competent mechanic demanded approximately the amount we had paid for the engine, plus half our foul-weather clothes, for his services. We gave him one-tenth of this, and he left, apparently moderately satisfied.

We then had the problem of the old engine. The customs officers informed us that not only could we not sell it in Greece, we could not leave it in Greece. It would have to be taken with us, and they would make sure that it was. There is little enough room on board *Tulip* and a second engine, which did not work, was more than we wished to carry. Our small cockpit was filled with a totally useless hunk of metal.

We were prevented from leaving that evening, as the harbor police had seized our documents and passed them on to the customs officer. He would not release them until regular office hours, when officials could properly supervise the departure of both engines. During this enforced delay at least I could clean up. I was filthy with engine oil and rust, so I asked in turn at ten of the hotels of Patras for a bath or shower. But none wanted anything to do with a scruffy, dirty, long-haired sailor, and I ended up washing in a bucket on the wharf.

The next morning, a cold gray one, the bureaucrats finally gave up and let us leave. When we were well out to sea, Jonathan and I heaved the old engine overboard, while Pieter photographed the event for posterity. The satisfying finality of the act was lost in our exhaustion.

"But oh! shipmates! on the starboard hand of every woe, there is a sure delight."[1] These comforting words were spoken by Father Mapple in *Moby-Dick* during his sermon to the departing whalers Queequeg and Ishmael. And sure enough, "Dolphins at the bow," Jonathan calls. The rest of us tumble on deck and struggle forward. Six saddleback dolphins (*Delphinus delphis*) crisscross beneath us, gliding through the steep waves, over which we labor, leaping clear of the streaming foam and flying spray. In these conditions we can do little more than mark the dolphins' presence. It is too rough to hold a camera steady or record sounds through a hydrophone. But they remind us of our goal — the sperm whale.

Feared, marveled at and hunted mercilessly, the sperm whale has sunk deep into the consciousness of man. More numerous and widespread than any other large whale, the sperm whale populations constitute a vital element in the ecology of the ocean. As they have in the economy of man: the Quakers of Nantucket who sent their sailing ships around the world in pursuit of sperm whale oil were the Texaco of the early nineteenth century. Sperm whales, with their huge casklike foreheads filled with oil, their unlikely separation of the polar males from the tropical females, are so different from any other animal on earth or in the sea that scientists have long pondered them. But despite the sperm whale's ecological, economic and biological significance, its image in literature is what has directly touched most humans.

In the great allegorical novel *Moby-Dick*, Herman Melville, a whaler himself, describes the sperm whale as a fierce, proud and transcendent being: " . . . declaring the Sperm Whale not only to be a consternation to every other creature in the sea, but also to be so incredibly ferocious as continually to be athirst for human blood."[2] " . . . touching the great inherent dignity and sublimity of the sperm whale . . . ."[3] "I then testified of the whale, pronouncing him the most devout of all beings."[4]

Who is this leviathan? Whalers, ancient and modern, have killed hundreds of thousands of them, but still we must declare with Ishmael, "I know him not, and never will."[5]

2

THE RED SEA

December 2 - 12

WE ARE AT SEA, SURROUNDED BY THE comforting rhythms and rituals of the ocean. The spinnaker Wuff-Wuff, which is drawing us gently over the smooth waters, is backlit by the sun, shining strongly through the billowing white fabric. Enclosed in our own small home, we are cradled by the swells. The ocean and an occasional passing ship form the outside world. The wind is Beaufort Force 3 (ten miles an hour) — perfect for effortless sailing. If it were lighter, we would have to coax her along, watching and adjusting the sails and the course, or even, reluctantly, starting the engine. In stronger winds *Tulip* begins to heel, pitch and roll.

"Clarence," our self-steering gear, is at the helm, and the crew can relax or work at preparing the boat for the research ahead. Clarence is perversely named after a drunken one-eyed Newfoundland fisherman who would remove his glass eye and hand it to you in order to make a debating point. But human foibles do not affect Clarence the self-steering gear, who keeps a straight course as long as there is a breeze to direct him, leaving the crew free from the helm.

Jonathan, our handyman, is sitting in the cockpit, making a small hammock for storing vegetables. Pieter, braced against the shrouds that support *Tulip*'s mast, is photographing Wuff-Wuff for the people back in Holland.

Pieter was given a potted plant (species unknown) by a Dutch friend before he left. It was installed in the place of honor at the center of our saloon and has just started to blossom. Its delicate smell drifts through the main hatch beside me, accompanied by hints of another fine meal from Nic.

A few days behind us is Egypt and the Suez Canal. It lay, a physical and bureaucratic block, between *Tulip* and the Indian Ocean. The recent assassination of President Anwar Sadat threatened the stability of a troubled region and the availability of the canal. But happily the political situation remains calm, and we had only to face the bureaucracy.

There were many large ships anchored off Port Said as we neared the Mediterranean entrance of the Suez Canal. We called the pilot boat on our VHF radio, and they asked us to come alongside. We did. The pilot boat was a rusty tug.

"Write name of boat and registry on piece of paper," they called to us.

We obliged and handed it over, wondering why the information could not have been conveyed over the radio. It soon became clear.

"And our present?" the "pilot" called.

"Huh???" We were perplexed.

"Where is our present?"

"Your present????"

"Where is our present?" he repeated.

"Oh, yes — your present . . . " we replied, beginning to catch on. Scared that if we did the wrong thing we would not be allowed through the canal, I produced our only bottle of whiskey.

The man's eyes lit up. "Our present!"

We passed it over, and in return received a wave in the

general direction of Port Said. We had reached the East.

During the next few days we were to become well aquainted with the Egyptian system of "presents." Most services were performed free or at nominal cost, but substantial "presents" were sometimes required. The most desirable kinds of "present" were U.S. dollars or American cigarettes, neither of which we had on board, so we were reduced to handing out our supplies of alcohol, chocolate and Greek currency.

For two days in Port Said I went from office to office, filling in forms, getting signatures and paying out various sums. However, the operation was considerably simplified by the Suez Canal Authority, who thoughtfully provided a list of offices to be visited and a map showing their location. The Egyptian bureaucrats were understanding. There were many smiles and cups of coffee as I waited in dilapidated high-ceilinged offices for forms to be signed.

For all of us Egypt was the start of an adventure — our first step outside the officially Christian world; our first view of the effects of recent war and the unease of an ever-present threat. There were bullet-pitted, shell-blasted buildings, sandbagged gun positions in the towns as symbols of a troubled past and an uncertain future. Calls to prayer rang from the mosques and many of the women were swathed from head to ankle. We had left our familiar cozy Western world.

The passage through the canal took two days. We spent the intervening night anchored in a small lake. A little before sundown two ragged soldiers, armed with guns, rowed a small wooden rowboat up to *Tulip* and tied astern, where they spent the night. We were not sure whether their purpose was to guard us or to prevent us from engaging in some unspecified illegal act.

However, in the morning, our tired-looking guards said, "Please give us presents."

This demand may or may not have implied they had rendered us some service.

Our "brand-new" engine spluttered as we neared the end of the canal, and stopped as *Tulip* tied alongside in Suez. Diesels seemed to be haunting me. Despair and frustration welled up. As I forced my body to dismantle the engine, I wondered if it would ever run again. After several unsuccessful probes into the nature of the problem and many hours of battle, I replaced the "brand-new" fuel pump with one Jonathan had stripped from our old deep-sixed engine. It ran! We sailed out into the Red Sea.

Ships pass, but there is no other sign of the land where humans reign, although less than a hundred miles away on either hand are countries so foreign it is hard to conceive while on this friendly, smooth sea — Arabia and Africa, bypassed. We have been told tales of pirates, wars and unbending bureaucracies in these parts. It is best to stay in the shipping lanes, near the middle of the Red Sea, neglecting the deserts, camels and coral reefs.

The Red Sea is a deep trench separating the continents. Perhaps it is sperm whale territory. It was listed as one of "the favourite places of his resort" by the early-nineteenth-century authority on sperm whales, Thomas Beale.[1] We keep watch, but will not linger here, even if we do sight our goal, the sperm whale. The dangerous, deep blue Red Sea is no place to tarry.

The sperm whale is the most geographically widespread of the great whales, and has been found almost everywhere that there is deep water. Only the shallow shelves bordering the continents, where most large whale species feed on small fish or plankton, are not for the sperm whale. Melville accounted him "no common, shallow being . . . He is both ponderous and profound."[2]

Beneath me lie two miles of water, a realm of darkness and immense pressures, more foreign than the moon. Down there are fantastic angler fish with long, luminescent lures dangling over their mouths, giant squid perhaps larger than the sperm whale itself, and many animals never seen by man, forming an intricate unfathomed mosaic. The sperm whales are nearly unique among these strange animals in having to return regularly to the surface to breathe. Perhaps they can bring us some intimation of this unexplored world.

Sperm whales are not spread evenly over the deeps. As he traced the imagined migrations of Moby-Dick, Ahab knew well that there are areas where sperm whales are scarce, and other areas, called "grounds," where at certain times of year they are common. The British, the first to circle the world in search of sperm whale oil, discovered many of the grounds during the late eighteenth and early nineteenth centuries. They were soon overtaken by Americans, whose stubby whalers explored almost every part of every ocean, leaving no sanctuary for the sperm whale.

There are sperm whaling grounds in all oceans, and at all latitudes except at the frozen poles. Mathew Maury, a lieutenant in the U.S. Navy, was the first to scientifically document the grounds. In 1852 he plotted the positions of sperm whale catches by American whalers.[3] Eighty-three years later Charles H. Townsend, the director of the New York Aquarium, published a comprehensive analysis of the positions in which the nineteenth-century Yankee whalers made their kills, the famous "Townsend's charts."[4]

These maps, together with sightings from British and Dutch merchant ships, helped us to plan *Tulip's* route, but last September Jonathan and I were lucky to meet Captain Mörzer Bruyns, a Dutch sea captain. Captain Mörzer Bruyns had kept a careful record of the whales and dolphins he had seen during a lifetime at sea. He is the world authority on where to go to find whales, especially in such an unstudied area as the northern Indian Ocean. In his *Field Guide of Whales and Dolphins* Captain Mörzer Bruyns describes animals he has seen regularly at sea but have yet to be otherwise described by scientists.[5] Perhaps those small dolphins that rode our bow last night were his "Red Sea Dolphins." Captain Mörzer Bruyns kindly allowed us to copy his meticulous sighting charts for the area, and they give us our best idea of what we may hope to see.

After emerging from the Red Sea, we plan to follow the course set by the Yankee whalers, who traveled with the trade winds, or monsoons as they are called in the northern Indian Ocean, to the Arabian Grounds off South Yemen and Oman, to the Colombo Grounds south of India, and finally across the equator to the Seychelles Grounds.

It is rapidly growing hotter. In Greece it was winter, in Egypt autumn, but now it is definitely summer, and each day is more torrid and humid than the previous one. At this time of year winds blow from north and south into the center of the Red Sea, which becomes a trap of moist heat.

Jonathan looks up, checks the compass and his watch and goes below. He brings a clipboard back on deck. It holds our Environmental Sheet, on which he records *Tulip*'s position and speed, the wind direction and strength, wave height, temperature and other environmental variables. These environmental sheets will trace *Tulip*'s wanderings, and the conditions she encounters, over the next three years. Jonathan also fills in the ship's log, and calls to Pieter to take over the watch.

"Anything to do?" Pieter asks.

"Just watch Wuff-Wuff," Jonathan replies. Wuff-Wuff is a temperamental sail, and needs regular attention if it is to keep full and drawing.

We each take four-hour watches; the watches become an important harmonic in our lives. We also rotate cooking the evening meal. In contrast to the stormy sickening days off Greece, these conditions give everyone a healthy appetite; meals have become the prime events of the day.

The fine weather not only allows us to relax our bodies, repair our equipment and eat with pleasure, it also gives us an opportunity to talk and to learn a little of what has led us here.

Pieter has the most to tell. He is older, comes from another country and has led an adventurous life. He established Greenpeace-Netherlands and has been in many parts of the world protesting against environmental destruction. Each protest, if successful, seems to end in the arrest of the protesters, making for widespread publicity, the key to success for these activists. So Pieter has had brief spells in prisons from Iceland to Peru, and has obviously enjoyed his vigorous life. Life on board *Tulip* must seem placid to him, and I hope he can find the "publicity value" that W.W.F. needs in our measured passage to the whales.

We have left the familiar Atlantic, whose whales I had grown to know, and Newfoundland, my Eldorado for so long, where there is work to do, a job and a home. Why? My rationale is similar to Ishmael's when he joined Captain Ahab on the *Pequod*: "Chief among these motives was the overwhelming idea of the great whale himself. Such a portentous and mysterious monster roused all my curiosity. Then the wild and distant seas where he rolled his island bulk."[6]

But not all share my love for the waters. Ishmael had no ease on "this appalling ocean," and longed for the "green, gentle, and most docile earth."[7] I talk with Nic and find her, too, less than wholehearted about the sea. "It can be nice and it makes a pleasant change for a bit," she says, "but I'm never sure what is going to happen next. I feel much more at home on land."

I do not yet feel close to this crew. They show none of the fellowship of my old shipmates. *Tulip* is quiet, bar the creak of a block. There are no recently composed raunchy ballads telling of our research among the fishermen of Bay de Verde, Newfoundland, no bad jokes about the curious habits of the skipper and no gentle love songs telling of the separations the sea invariably forces.

A long smooth swell flows from the south, presaging some rugged times ahead, but for now, there is peace on the ocean.

3

RED SEA TO KURIA MARIA ISLANDS, OMAN

December 13 - January 18

NONE OF THE OBSTACLES IN *TULIP'S* PATH had been as fierce as the appropriately named straits of Bab al-Mandab, Arabic for "The Gate of Tears." None had seemed so cruel, so powerful, so determined to bar us from the Indian Ocean and the whales.

As *Tulip* made her way southward through the Red Sea, the favorable winds, which had kept Wuff-Wuff billowing and our progress excellent, gradually died. Light, shifty southeasterlies forced us to tack back and forth across the Red Sea. However, we tried to avoid the coast of Ethiopia, where the Eritrean War was being fought, and we had heard of yachts being shelled. As *Tulip* sailed south the winds grew stronger, until, on approaching Bab al-Mandab, through whose narrow gullet the Red Sea is joined to the Gulf of Aden and the Indian Ocean, we were beating against a full gale. The winds pouring into the low-pressure area in the center of the Red Sea gained strength as they were funneled through Bab al-Mandab.

When astern, gale force winds may be exhilarating and can

make for fast progress, as the boat lifts and surfs, riding with the strength of the blow. But to beat against a full gale is to struggle against all the power of nature. The course can be no closer than forty-five degrees to the direction of travel, and a wind-driven adverse current slows any progress to a crawl. Each wave slams against the hull, stopping the boat and churning the stomachs of the crew. The water pouring over the decks forces all hatches to be closed and makes its way through small cracks in even the soundest hulls. Cracked containers, dislodged fruit, or broken eggs add to the putrid potion that, as the boat heels far to leeward, makes its way from the bilges up into lockers and bunks. The cabin becomes a fetid, stifling, squalid hellhole.

As *Tulip* beat her way to Bab al-Mandab, two fuel jerricans split, so we had diesel oil sloshing over the cabin floor. The fumes were nauseating, and as the slick diesel washed over the floorboards, there was now no secure foothold down below.

Then, on December 20, a particularly strong gust blew the jib to tatters. We anchored behind Little Hamil, a volcanic ash heap of an island. There we had shelter at least from the waves, but the wind continued to scream. We cleared up the diesel, and thanks to Nic, who so often in the worst situations would take care of our weary bodies, we sat down to a fine meal.

But our situation was utterly frustrating. There were only eighty-five more miles of the Red Sea before the Indian Ocean, where we could expect much more pleasant weather. Without the jib it would be hard to beat through it, and with our two split jerricans we no longer had enough fuel to motor the whole way. We could wait, but these winds might last for weeks.

I decided to fight. "Let's go."

Jonathan, who is more careful, did not agree. "It's stupid. Our equipment is likely to be damaged. We should wait for better weather."

But the captain always has the last say, and despite Jonathan's misgivings, we left our anchorage, raised storm jib and fully reefed mainsail and beat on into the gale. But a few hours later the wind had strengthened enough so that we had to lower the last patch of mainsail. *Tulip* would not beat under the tiny storm jib alone, so we turned on the engine and used our last fuel to motor to the Yemeni port of Mocha. The gale had now picked up sand, and we needed goggles to see the course ahead. The engine, so unreliable at other times, carried through, and we tied alongside late on December 21. We needed to buy ten gallons of diesel, which ought to be enough to bring us through Bab al-Mandab to Djibouti.

Mocha, which has given its name to the fine chocolate-flavored coffee that used to be grown in the area, was a desolate place in the sandstorm. Although a small modern port had recently been built by the Dutch, the town was little more than rubble following the recent troubles between Yemen and the neighboring, ideologically opposed, country of South Yemen. The local bureaucrats, wearing skirts and daggers, were generally helpful, and our stay culminated with my being whisked off across the desert, past camels and gun emplacements, on the back of a motorbike, clutching two jerricans.

The following day, once again with Jonathan's grave misgivings — "Hal, I really think you are crazy" — we motored out into the teeth of the gale. For twelve hours we pounded through Bab al-Mandab, dodging the shipping that is funneled through these narrow waters. Finally *Tulip* emerged into the gentler climate of the Indian Ocean. I was exhausted, shattered. Before leaving the watch to Pieter and falling asleep, I pondered why I push so hard, so stubbornly. Was it necessary? What drove me? But we had succeeded.

Despite the dire warning in the British Admiralty pilot book — "Djibouti is one of the hottest and grimmest parts of Africa"[1] — Djibouti proved to be a fine place to spend Christmas. A very small country, it used to be a French colony. There is still a sizable French military presence; *légionnaires* and *matelots* dominate one part of the capital and principal port, which is also called Djibouti. However, a few hundred yards from the expensive cafés and shops with Christmas trees in the windows, refugees from the Ethiopian civil war sleep on the streets.

In this strange concoction of urbane France, loosely stirred with the African and Arab worlds, we surveyed the damage that Bab al-Mandab had wreaked on *Tulip*. The VHF radio aerial had blown away, as had Clarence's wooden wind vane, but the tattered jib was our main concern. There was no commercial sail-maker in Djibouti, but on board a French minesweeper support vessel we found Yves Graill, a sail-maker from Quimper, Brittany, *Tulip's* last permanent port. Yves was carrying out his military service without much joy, and seemed only too happy to spend his spare time on *Tulip*, reconstructing the jib. For three days we sewed away under his tutelage, and at the end *Tulip* had a serviceable, if not particularly elegant, jib. Yves would take nothing except meals and thanks for all his work, but for us the jib is now "Yves's sail."

Gay Alling, our final "permanent" crew member, flew in from New York early on Christmas morning. A student at Middlebury College in Vermont, Gay had been writing to me consistently for several years, asking to help with our whale research. She was obviously determined, but I always wrote back: "Whenever possible, I prefer to work with those I know personally. But if I ever need someone, I will let you know." Finally, last summer I needed one more person for our study in Greenland, so I invited her to come along.

Gay is small, slim, with long blond hair, pale skin and fine strong features. Originally from New York, she is also graceful, athletic and immensely determined.

Greenland proved much more difficult than we had imagined. The weather was appalling, the navigation treacherous, and our boat needed constant maintenance, especially the engine, which was even more troublesome than *Tulip's*. One crew member found out that she was pregnant. This interacted poorly with the motion of the sea. Sadly, she returned home. Another could not take the life, and three of us were left to carry on. We were cold, exhausted and wrought up for most of the time. It was largely Gay's determination that pushed us on, her warmth that kept us stable, and it is to her credit that the research itself was successful. I was surprised, but most pleased, when, having endured the most difficult whale study that I had ever known, she agreed to join me again in the Indian Ocean.

Gay brought with her plum pudding, liqueur and other goodies. We spent Christmas day snorkeling among some coral islands off Djibouti. On the way out *Tulip* passed the small gray humpbacked dolphins, *Sousas*, who could sometimes be seen puffing around Djibouti Harbor. To complete a memorably different Christmas, Jonathan was able to swim with a pair of mating turtles.

Despite the opportunity to rest, walk and glimpse a new continent, we were all happy to return to sea on December 29. And now, with a full crew, our equipment tested and rigged and the rigors of the Red Sea astern, we were ready for the whales.

The mood of the crew had improved considerably with the Christmastime recuperation and the injection of Gay's energy and enthusiasm. Jonathan has an inquisitive naturalist's eye for the life around him: he began collecting plankton, watching seabirds, which are relatively plentiful in this productive region, and identifying the turtles we encountered. Nic at last seemed to feel some joy in what we were trying to do, and Pieter was ready to film it all, especially sperm whales — if we could find them.

But despite assiduous watching and regular listening through the hydrophones, we neither saw nor heard sperm whales on our slow zigzag out of the Gulf of Aden. There were two or three distant puffs, which, despite vigorous chases, became nothing. There were some small, strange beaked whales, probably of the suitably exotic-sounding *Mesoplodon* genus, but none of their larger cousins.

The productivity of the Gulf of Aden and the Arabian coast was delightfully demonstrated by the numbers and variety of dolphins we saw. At night they came to *Tulip*. We sat on the pulpit at her bow and watched their weaving phosphorescent tracks teasing *Tulip's* sturdy bow wave or bursting from the water like fireworks. We heard their squeals, pitched at the upper limit of our hearing, the faint puffs of their blows and the splashes as they reentered the water. They came singly, or in groups of up to fifteen, staying with us for thirty seconds or two hours.

During the daytime we could watch their natural behavior away from *Tulip*. We tracked them, recorded their sounds, photographed their leaps and tried to work out which species we were seeing. This was Gay's work — the dolphins we sight are her project — and she invested a large amount of energy in her graceful friends, despite the exhausting trial of being woken every hour or two throughout the night by the watchkeeper: "Gay, wake up, there are dolphins at the bow. Small, maybe *Stenella*."

Some were species with which we were familiar. There were lithe saddlebacks, *Delphinus delphis*, sometimes called the common dolphin, their characteristic crisscrossed flank pattern briefly visible as they leaped through the ocean in vast schools. Saddlebacks have a worldwide distribution, but there are variants, and these animals seemed a little larger and lighter colored than those we saw in the Mediterranean.

The large, light gray Risso's dolphins, *Grampus griseus*, became familiar to us as we sailed east along the Arabian coast. We often saw the Risso's, with their scimitarlike fins rolling from the water, cruising the edge of the continental shelf. On climbing the mast, one could see the pale scarred bodies of the older adults. The Risso's must go through some tough times to gain such impressive scarring, but the animals that we saw seemed gentle and were generally undemonstrative; only rarely did they leap or porpoise, or approach *Tulip* closely.

Most endearing were the small spinner dolphins, *Stenella longirostris*, cavorting in large schools, displaying their spectacular leaps. Apart from the characteristic spinning jumps, in which the animal leaves the water spinning about its longitudinal axis, they would do cartwheels, head-over-tail flips and a whole gymnastic routine of aerial behavior. This activity does not seem to be concerned with feeding: with their long snouts (*longirostris* translated) the spinners generally eat only small fish and squid. Leaping is probably most often a social signal.

There were other dolphins, which we photographed in the hope that they may later be identified by experts. They were unfamiliar to us; on some occasions, despite excellent views of animals riding *Tulip*'s bow wave, the dolphins we saw seemed to match none of the descriptions in our field guides. They may have been subspecies, or even species that were new to science.

We sailed along the Arabian coast to Ras (Cape) Fartak in South Yemen and Ras Marbat in Oman, two of the Yankee whalers' favorite haunts — but no whales. Were we too late in the year? The Yankee whalers had usually arrived one to two months earlier. Had the whales moved their grounds during the intervening century? Or had the modern whalers finished their predecessors' bloody work? It was a disappointment that not even all the delights of the dolphins could overcome.

On January 10, 1982, we put into Salalah, Oman, yet to see a sperm whale. We needed to make a few repairs, replenish our supplies, and it was here that Pieter would leave us, to be replaced for the next two weeks by an Omani. A young slim, dark man ceremonially dressed in a turban and a dishdash, the long white tunic of the Omanis, greeted us at the wharf and handed over an official-looking letter from the Ministry of Agriculture and Fisheries of the Sultanate of Oman: "This to welcome you to Omani waters and to introduce you to Khamis Suleiman al-Farsi, who will join you as counterpart for the rest of your trip to Muscat."

"Would you like to come on board?" we asked.

"Of course," he replied in excellent English.

Boarding a small boat is not an easy task when wearing a dishdash, but he managed it with remarkable agility. Was this to be our Omani crew member? Life on a small, congested yacht with women, vegetarian food and (a little) alcohol did not seem to be obviously compatible with the traditional Muslim lifestyle. How would he move around at sea in that strange costume? Where would he spread his prayer mat?

I tried to broach the subject carefully. "Would you be at all interested in joining us during our studies? As you can see, it is cramped."

"Of course."

He seemed offended there had been any doubt. Well, it would only be for two weeks, and maybe he would adapt.

At Salalah we had a glimpse of this newly oil-rich Arab state. Mostly we were confined to the fine, clean, modern port, run almost entirely by South Asian expertise and labor. The Arabian oil states depend greatly on workers from India, Pakistan and Sri Lanka, who literally "make their fortunes" during a few years on the Arabian Peninsula. The port is virtually a prison for these foreign workers, but for us it had some attractions: humpbacked dolphins (*sousas*), which seem to be attracted to harbors in this part of the world, puffed their daily rounds beside *Tulip*. However, we longed to see the real Oman beyond the port.

Our counterpart, Khamis, and Mr. Ali, who was in charge of Fisheries in Salalah, took us into the town so that we could buy supplies, take showers and eat in a restaurant. Salalah, set on a coastal plain beneath sandy mountains, is one of the most productive areas of Oman. There were palm trees and other signs of cultivation, but our Western European eyes saw a dry, barren land. We found a half modern, half ancient world: street bazaars and Holiday Inns, camels and highways. Most memorably, one afternoon Khamis and Mr. Ali took us up into the mountains to a gorge where a spring allows the cultivation of a lovely garden in the desert. Here Pieter, Jonathan, Nic and I climbed with our hosts up a small path to some caves. Beneath us we witnessed a strange and delightful encounter, illustrating what is probably the greatest dichotomy between the Arab and

Western worlds, as Gay met and talked with a group of about fifteen young Muslim women. We could not hear what was said, but we watched entranced as the steady blond American woman in her light green print dress laughed with the willowy Arab ladies in their black gowns and veils. Her emancipated way of life was as unimaginable to them as their shuttered existences were to her.

On January 13 we put to sea again. Our worries about Khamis were baseless; he climbed on board *Tulip* wearing Levi's and a Western shirt, with almost no other belongings. He had spent several years studying in Aberdeen, Scotland, and was familiar with boats, having often worked as an observer on foreign vessels fishing off the Omani coast. He knew these, his waters, well, and taught us. "You see the whales just here," he said, pointing to areas on the chart, "and in this bay here is where the fishing boats worked." Khamis was easygoing and had a fine, dry sense of humor. He also had a sharp eye for whale blows.

Yesterday, beneath the misty barren mountains of Jabal Nus, off the coast of Arabia, we found our first sperm whales. It was a bright, rough afternoon, with dolphins everywhere. We tried to photograph them as they leaped between the waves. Suddenly we saw what was not a dolphin, but a low bushy cloud of mist.

"A blow!" I cried. "It's a whale."

Another blow. This time we were watching intently. The blow seemed slanted, at about forty-five degrees to the horizon.

"It looks like . . ."

Another.

"It is — a sperm!"

After about fifteen more blows, spaced about twenty seconds apart, the smooth triangular flukes, or tail, rose to the sky as the whale began its journey to the deep. To Melville this was "perhaps the grandest sight to be seen in all animated nature. Out of the bottomless profundities the gigantic tail seems spasmodically snatching at the highest heaven."[2]

We did not immediately sense the metaphysical nature of the sperm whale's flukes, but we were filled with the simple joy of realizing our first goal.

The two months since we had left Greece, and the trials they contained, had aimed our lives at this moment. We had prepared for it, longed for it, in some ways even dreaded it. If we found the sperm whales but could not study them, what then? And as the weeks wore on, we had begun to wonder if we might never see them.

Through the headphones I hear a rhythmic knocking, the sound of success. The headphone cable leads through amplifiers to a hydrophone (underwater microphone) suspended over *Tulip*'s stern, and it is picking up the sounds of a sperm whale hundreds of feet beneath.

The knocking is almost metronomic at about twice per second — a muffled hammering that gained the sperm whale an unlikely alias: "the carpenter fish." But as I listen, the knocking pauses for two seconds, restarts at a faster rate and then accelerates, turning into a lingering creak. Another pause, this time for five seconds, and then the regular knocking recommences. What has happened so far beneath us?

Three minutes later the carpenter becomes a horse at a brisk trot as two other sperm whales add their reverberant clicks. Each whale clicks at about the same rate (twice per second), but there is no indication of synchronization. Several more whales join in, and now the chorus sounds like a horse race on hard ground.

As the clicks pause, start again, pause and finally halt, it is time to watch. Our short experience of sperm whales suggests that they will soon be at the surface. We scan the horizon, each taking a ninety-degree arc. About two minutes later Khamis spots the low bushy blows four hundred yards dead to windward. We call this pair a "cluster," meaning whales seen swimming together at the surface, but not implying any long-term relationship. I turn the engine on as Jonathan lowers the jib. As we motor up to the whales they startle, arching their backs to dive. We stop and scan the horizon once again.

"There they are, about two hundred and fifty yards away," calls Jonathan after about four minutes. This time I approach more gently, and the whales are less skittish. As we come up to within thirty yards I put the engine in neutral and we coast along beside them.

The whales lie at the surface like two wrinkled logs, each about thirty-two feet long. They are moving forward slowly at about one knot (a little over a mile an hour), but if it were not for the blows, which appear like windblown bushes from the whale's forehead, it would be difficult to tell which end was which. The sperm whale blow, for intricate anatomical reasons, is slanted forward and to the left at forty-five degrees to the vertical.

The two whales are almost inanimate in their quiescence. Their dorsal fins, extrusions of the rough dorsal ridge, bob up and down. One fin has a diagonal scar across it, a cheering sign: we hope to be able to identify individual sperm whales from such marks. With this in mind, Gay photographs the fin.

The whales are breathing about once every fifteen seconds, and after about forty blows each, they accelerate, raise their square foreheads two feet out of the water, blow and roll their backs to the strong downbeat of the flukes. Another blow, and the gnarled backs arch gothically — a most unexpected action from the lethargic logs we were watching a minute ago. One of them follows by throwing its flukes into the air. The other is less ostentatious, slipping modestly beneath the waves. Gay photographs the flukes, again hoping for distinctive marks that might identify individuals. However, it is unlikely that we will find many as easy to recognize as Moby-Dick, whose "broad fins are bored and scalloped out like a lost sheep's ear!"[3]

The water in the "footprint" left by the diving whales turns a greenish brown. "Shit!" I yell excitedly. Even though I am not in the habit of swearing with quite so much enthusiasm, the others assume that the mainsheet has tangled, or that I have suffered some similar minor catastrophe, and continue to write notes or check the film in their cameras.

"Shit! Shit! Shit!" I grow more excited, steering *Tulip* toward the discoloration.

Jonathan, who is quickest to realize what has happened, grabs the plankton net and scoops at the fast-disappearing feces. In the net are two small translucent pyramids, each about half an inch long and resembling the beak of a parrot. These are in fact the beaks of squid, remnants of the whales' food. From these beaks the species and approximate size of the original squid can be deduced. This means that sperm whales do not need to be killed with exploding harpoons, dragged to shore or a factory ship and then cut open to investigate their feeding habits; one of the scientific justifications of whaling begins to founder. It is an exciting moment: *Tulip* has made one small step on the path to her principal goal.

Having collected the squid beaks, we listen for the whales through the hydrophone. Sure enough, the knocking clicks start again a few minutes after the dive.

"If we could only track them . . ." I think wistfully.

As I left Newfoundland for Greece and the start of this voyage, my colleague Dr. Jon Lien gave me a strange-looking object: a directional hydrophone. It consists of a cone, about nine inches in diameter, made of a metal alloy, but covered with neoprene rubber. Underwater, the neoprene reflects sounds entering the cone to a hydrophone placed along its axis. It was built for tracking small acoustically tagged fish in a tank, but it might be what we need. We lash it to the end of the boat hook, connect it to the acoustic system, and to our great satisfaction, we have a directional hydrophone that can be used to follow sperm whales.

It is, not surprisingly, rather difficult to use. The operator sits with legs over the side of the boat, bracing the boat hook between feet and hands, as the hydrophone is dragged by every wave and surge. Headphones allow us to hear the sounds, and the outfit is topped off with a towel draped over the ears to help block extraneous sounds. Despite the ungainly pose, we find that, with sufficient twiddling of fingers and toes, we can usually obtain a rough bearing on the clicks, and thus we can follow sperm whales! This is so important to our work.

As I ponder what the whales might be doing beneath us, a series of large gray ships start appearing from over the southern horizon. The fifth is immense, and as it steams closer, we can make out an aircraft carrier. This is part of the U.S. Navy's Indian Ocean Fleet. Soon jet fighters are wheeling a few yards above *Tulip*'s masthead. Are we a threat to American security? It is an incredibly brazen intrusion on the life of the sea. These behemoths have sailed halfway around the world to sully this sanctuary.

Despite the distraction, we manage to keep following the whales, until there are suddenly loud metallic pings, distinct from the dull knocks of the sperm whales, coming through the hydrophone. Some work of man is down there. Is it the sonar of a submarine, or perhaps a sonar searching for a submarine? The sperm whales are silent.

We listen for an hour as the sun starts to set behind the desolate Omani mountains, but no clicks. There is, however, a very faint moaning. It stirs a familiar chord . . . . Could it be? Perhaps?

We turn *Tulip*'s head and sail her north for about ten miles, over the edge of the continental shelf into Kuria Maria Bay. As we sail, Gay cooks up a fine dinner. Through us all resonates the excitement of what we have experienced, and anticipation of what may be about to happen.

On reaching shallow water the hydrophones are lowered. To cries of "Yes, it is! Oh! Ye-e-s," pealing through the boat is the song of a humpback whale. The extraordinary series of moans, grunts and whistles are familiar to me from winters spent on Silver Bank in the West Indies, the humpbacks' major North

Atlantic wintering area. But the others have only heard the songs on the recordings made by Roger Payne.

The humpback whale is a strange, knobbly creature, about forty feet in length, with long armlike flippers and an engaging disposition. Herman Melville called it "the most gamesome and light-hearted of all the whales, making more gay foam and white water generally than any other of them."[4] And Frank Bullen concurred: "There be few creatures in earth, air, or sea, that lead a happier life or enjoy it with more zest than the humpback."[5]

Slow, coastal, predictable, demonstrative and lovable, the humpback has become a favorite of both the whale-watching public and scientists developing benign research techniques.

Many have tried to describe the song of the humpback, but it is more than "sonorous groans" or "unearthly wails." They use the highest notes we can hear, the lowest, and all in between, to construct an eloquent adagio. The songs were discovered and analyzed by Roger Payne and Scott McVay, whose paper in the journal *Science* is one of the classics of modern whale research.[6] The song is made up of about eight *themes*, each containing several identical *phrases*. Each phrase is made up of single sounds, called *units*. The whale sings the phrases of *Theme I* any number of times, then moves to *Theme II*, which also can go on for as long as the whale wishes, and then to *Theme III* and so on, until eventually, after about ten to twenty minutes, it returns to *Theme I*. There is no official beginning or end, although a ratchetlike theme usually coincides with the whale's breathing. The whale may sing just a few *phrases*, or go on for hours. As the series of sounds is both structured and cyclical, it is a true song.

Recent studies by Jim Darling, Peter Tyack and others on the humpback whales off Hawaii, as well as our own work on Silver Bank, have begun to shed light on the function of the humpback song. The whales very rarely sing in the cold waters where they feed in summer; instead we hear their songs during winter, when they are mating and calving over shallow banks in the tropics. Singing humpbacks are almost always single males, and Peter Tyack suggests that singing somehow assists the lonely bachelor's chances of joining, then mating with, a female[7]; the song is probably a form of courtship melody.

Roger and Katy Payne have made further remarkable discoveries. In any singing area (such as Silver Bank or off Hawaii), at any time, all the humpbacks are singing more or less the same song, but during the course of a four-month winter season, one theme may evolve, another drop out, so that by the time the whales are ready to return to their summer feeding grounds, they are singing a rather different song. So there is an ever-changing "top tune," which all sing. Over the years the changes build up, so that songs recorded in Bermuda by the Paynes ten years apart are totally different in content, although they follow the same set of rules. Likewise songs of the same season from Hawaii in the Pacific and the West Indies in the Atlantic have different components, despite a similar structure.[8,9]

This song from Kuria Maria Bay — as far as we know, the first recorded in the Indian Ocean — although made up of *phrases*, *themes* and *units*, contains sounds unlike any I have heard before.

The songs tell us so much: first, that there is a lonely humpback lurking somewhere nearby on this dark Arabian night; also, that this is his mating ground and that he thinks it is winter and time for mating. This might not seem surprising, since we are in the Northern Hemisphere and it is winter. But there are no cold waters in the northern Indian Ocean where the humpbacks might feed. Other scientists receiving reports of isolated humpbacks off Arabia, or in other parts of the northern Indian Ocean, assumed they were stragglers from the southern Indian Ocean populations that feed in the Antarctic.

But from the timing of his singing, this is a Northern Hemisphere animal. Then where does he feed? The waters off Oman become extraordinarily productive during the summer months, when they are churned by the powerful southwest monsoon. As Mr. Ali, the head of the Omani Department of Fisheries in

Salalah, told us, to the envy of his counterparts in other countries, "Too many feesh, too many feesh!"

Perhaps this humpback only needs to swim a few miles, rather than to a cold polar ocean, to find his summer's nutrition. If so, he would be spending his entire life within the Indian Ocean whale sanctuary.

We spend the night listening to and recording the song, occasionally breaking off to sail *Tulip* a mile this way or that in order to improve the quality of our recordings. Tomorrow morning we will head back to the deep waters for sperm whales.

4

OMAN TO GULF OF MANNAR, SRI LANKA

January 19 - February 27

OUR TIME AMONG THE WHALES OFF OMAN was limited: we spent four days with the sperm whales in Kuria Maria Bay and four nights inshore listening for, and sometimes recording, humpback songs, although we never saw a humpback. During the remainder of our passage to Muscat, near the eastern tip of Arabia, we were visited by a procession of dolphins; we saw seabirds, turtles and a rare Cuvier's beaked whale (*Ziphius cavirostris*), but none of the sperms that we principally sought.

Khamis left *Tulip* in Muscat, a spiky Hobbitland — rocky and mountainous, with fort-capped pinnacles of sandy rock. Muscat is the capital of Oman, and here we received the generous hospitality of Khamis's superior, Mohamed al-Barwani, the director of Fisheries Research, a splendidly energetic and cheerful Omani. Mohamed has represented Oman at the International Whaling Commission for the past few years, and takes justifiable pride in the support that his small but maritime nation has given to whale conservation. Mohamed has also been active within Oman, encouraging local ships to record whale sightings and assisting with the visits of foreign scientists such as the *Tulip* crew.

The Ministry of Fisheries in Muscat put on a special dinner in our honor. In preparation Gay and Nic bought me a new shirt, and Khamis drove us up to a spectacularly located luxury hotel. I was seated by the side of the minister of Agriculture and Fisheries. I tried to make conversation, but it was soon apparent that my neighbor's English, although putting my nonexistent Arabic to shame, was insufficient to discuss whales, or even fish, meaningfully. So I turned my attention to the food that was being set before us. It was delicious, and the *Tulip* crew ate heartily of the shellfish, humus, baba ganoush and other Middle Eastern specialties. This was in sharp contrast to our Omani hosts, who just politely nibbled. It soon became apparent why. As we began to feel pleasantly satiated, the waiters removed the food we had been eating and served newly laden plates. We had stuffed ourselves on the *hors d'oeuvres*! Course followed course, and although the actual contents of the later courses are blurred in my mind — there was lobster at some point — the *Tulip* crew were undaunted and we made complete pigs of ourselves, while the Omanis watched politely. They must have wondered if we ever ate while at sea.

We were sufficiently behind schedule that we rushed off from Muscat on January 27 after a brief three-day stay. As all reports suggested there would be few whales on the next part of the voyage to Colombo in Sri Lanka, we took no visitors on board for this leg of the journey. Our passage from Muscat to Colombo fit Melville's description: "For this part of the Indian Ocean through which we then were voyaging is not what whalemen call a lively ground; that is, it affords fewer glimpses of porpoises, dolphins, flying-fish, and other vivacious denizens of more strirring waters, than those off the Rio de la Plata, or the in-shore ground off Peru."[1]

It was a slow sail, with head winds during the first part and extensive calms later on, and as we sailed south the heat grew more oppressive. When on watch, we were constantly adjusting the helm and the sheets to maintain *Tulip*'s heading and speed. Off watch there was little to do but read, write letters, or listen to Jonathan's extensive collection of taped music. If the wind was very light we could cool down by diving over the side to experience the strange sensation of swimming in perfectly clear, immeasurably deep water — floating above the abyss. If *Tulip* had any forward progress, swimming was impossible, so we would wash and cool ourselves by tipping buckets of seawater over our heads.

Jonathan set some fishing lines using lures that Mohamed al-Barwani, another keen fisherman, had given him. We first caught a beautiful dorado fish, which was hard to kill but tasted excellent. It was followed by two tunas, each arriving on the day I, a fairly staunch vegetarian, was due to cook. The others helped prepare the fish, but only Jonathan enjoyed eating them. Finally we caught a large shark, which thankfully broke the line before we could haul it on board. After this, no one had much enthusiasm for fishing and the lines were put away.

We all felt depressed; my moods were as unsettled as Wuff-Wuff, sliding listlessly past the forestay, threatening a fatal wrap. I felt imprisoned by *Tulip*'s hull, the heat and my responsibility, and longed to run free on cold, rocky headlands, as had been my joy when portbound by bad weather in Newfoundland.

But later in the day the wind might pick up, a turtle pass, and I could sit against the mast in the shade of a well-rounded Wuff-Wuff. When I was lulled by the power of the ocean, peace would return.

Finally, on February 12 we rounded Cape Cormorin, the southern tip of India. It was early morning. The misty brown mountains of Kerala lay twelve miles to port, but around us the sea was dotted with small dark triangles, the sails of tiny fishing boats. As they sailed past *Tulip* on their way out to sea for a day's fishing, we could see that they were little more than logs rigged like Arab dhows, with crews of three or four. Their expectations cannot have been great, for there was little space on board for any fish that might be caught. There were several hundred of these boats, and they formed an elegiac introduction to the poverty, overpopulation and resourcefulness of southern Asia.

Later that day we passed over the edge of the continental shelf to the deep waters of the Gulf of Mannar, and almost immediately, for the first time since Oman, heard through our hydrophones the clicks of sperm whales. We were short on supplies so could not linger, but the sounds boded well for our time in these waters.

Tulip arrived in the large commercial harbor of Colombo, the capital of Sri Lanka, two evenings later. The next morning we walked into town to encounter the new, old world of Asia.

The port is a world of its own inside Colombo. Separated from the rest of the city, its labor force of several thousand live largely within its high walls. The port has its own dormitories, canteen and post office. Walking toward the gate we pass an ominously named "Lash Centre," whose function we have yet to determine. Curious Sri Lankan port workers stare or beckon to us. They keep wanting us to come and talk so that they can practice their English. The women crew members receive even more attention: the only woman we have seen working in the port is the Port Health Commission doctor. We pass a sawmill with a pile of huge hardwood tree trunks stacked outside, coveted by those of us who have worked with wooden boats. Farther along there is a wharf where Indian sailing ships are berthed, loading and unloading. We have seen these ships pursuing their ancient trade in rice, spices and other, more mysterious merchandise across the Gulf of Mannar, with sail piled on sail. The tangle of spars and lines of coolies hauling sacks reminds me of faded photographs of western seaports a hundred years ago.

On reaching the gate, we come to the customs checkpoint. We have learned, when carrying anything of value, to put on our most resolute expression and stride determinedly out, perhaps nodding to the customs officer. If he is in a good mood, we will receive a wide warm smile in return. A less well-disposed official will call us into his office, search our bags and, if he finds a camera or engine part, demand a one hundred rupee bribe. If we refuse, we have to walk right around the port to the main customs building and spend an hour filling in forms and getting them approved in order to temporarily carry the offending articles out of the port.

With the even-paced, dark and enclosed world of the port behind, we are assaulted by the noise, colors and smells of Colombo. There is a gaudy statue of the Buddha beneath a bo tree. A lady in a regal gold-trimmed sari and a man wearing a striking sarong swirling with vivid aquamarine briefly emerge before being reassimilated into the streaming throng — white shirts and dark skins. Shouts of street merchants, the screech of car horns and the permuting smells of burning palm oil, sewage and incense sweep through our senses.

Colombo, especially the area around the port, is not a particularly pleasant place. It is a hot, dirty mass of humanity. The few attractive buildings are out in expensive suburbs, and the canals and lakes that wind through central Colombo are reminders of a city built on a swamp, rather than the symbols of an aquatic civilization.

Despite most surface appearances, our feelings for Sri Lanka were generally favorable. The people, with their strong cheekbones, wide mouths and shining wavy hair, were friendly and usually spoke good English. To my relief, there was no sign of the militarism that was so rampant in the other countries we had visited. In contrast to Egypt, Yemen, Djibouti and Oman, there were no gun-toting soldiers at every gate and street corner. Instead Buddhist monks, clad in simple saffron robes and carrying umbrellas, strode serenely through the throng. Huge trees had been allowed to grow for centuries. Shrines were built around the trunks, and the branches formed vast canopies above the squalor of Colombo. Another contrast with the countries we had just left was the vegetation. Instead of barren deserts, there were lush tropical plants, which found their way even into corners of downtown Colombo.

We were constantly approached by street hawkers with an extraordinary variety of wares. With my long hair and scruffy appearance, I was a natural prey.

"Marijuana, sah?" a voice whispered distinctly, as I walked through the crowded streets.

"No, thank you," I replied, walking on.

Deciding that his pitch must have been too low, he went on. "Hasheesh, ve-ery good?"

"No," I said with a laugh.

The shadowy figure glided closer, and continued. "Opium?"

"No."

His last offer was a low piercing whisper. "Heroin?"

Immediately following my final "No," he disappeared into the crowd, doubtless pondering the vagaries of Westerners.

Jonathan's more civilized appearance and worried expression may have suggested he wished to return home, for he was approached with "Airline tickets, sah? Ve-ery cheap. Come with me."

Gay was obviously insufficiently adorned. "Jewelry, madam? Very fine stones."

And Nic, for no obvious reason, received the extraordinary proposition "Underpants, madam?"

We had been given contacts in Sri Lanka by World Wildlife Fund. Our first was Dr. Ranjen Fernando of the Wildlife and Nature Protection Society. Sri Lanka, partly on account of the pro-conservation ethic of the prevailing Buddhist religion, has an excellent record in terrestrial conservation, especially considering its poverty and overpopulation. If only Christ had taught his followers to respect all living things as clearly as the Buddha, how much more of the Western world might have been saved! In Sri Lanka wildlife is beautifully depicted on paper currency and postage stamps as a celebration of the Sri Lankans' lush surroundings. Over many years the Wildlife and Nature

Protection Society has been instrumental in the setting up of national parks and other conservation projects.

But marine conservation, as Dr. Fernando told us, was at a different stage. The Wildlife and Nature Protection Society had established some small sea-turtle hatcheries, but otherwise little was being done for sea life, and virtually nothing was known of the whales and dolphins off the coast. Dr. Fernando was surprised to hear about our whale sightings. Sri Lankan scientists, especially D.E.P. Deraniyagala, had written papers about the occasional whale that stranded along the Sri Lankan coast, but these strandings were assumed to be "accidentals." Twenty years ago Dr. Deraniyagala, in the Sri Lankan tradition of conservation, had actually proposed an Indian Ocean sanctuary for "turtles, the dugong, whales and dolphins," but his plea was buried in the scientific literature, and it is the Seychellois who take the credit for setting up the Indian Ocean whale sanctuary. The dugong is a herbivorous marine mammal that is generally found in shallow coastal areas of the tropics. In Sri Lanka, and throughout most of its range, the dugong is close to extinction.

Tulip left Colombo on February 18 for two weeks at sea. With us came Lex Hiby, a specialist on whale censusing. Although Dutch by birth, Lex works at the British government's Sea Mammals Research Unit. He is currently trying to work out how to census sperm whales most efficiently. He was initially interested in seeing whether our studies of whale behavior might be used to improve ship or airplane censuses. During these counts, whales are often missed because they are underwater or inactive. A knowledge of blowing rates, grouping patterns and diving intervals would allow the results of the censuses to be corrected. However, our methods of detecting and following sperm whales acoustically suggest that listening for their sounds may be the best way of counting them. Sperm whales are at the surface for about one-quarter to one-third of the day, during which time their blows and backs can be detected at a few hundred yards with reasonable reliability, between dawn and dusk and with good visibility. In contrast, they are clicking for over half the time, and can be heard through hydrophones at any time of day, in almost any weather, to ranges of several miles. And so Lex has now changed his emphasis from the visual to the acoustic, and is considering how hydrophones may be best used to census the whales.

With his humor, his rigorously inquiring mind and merry chuckle, Lex brought a sparkle into the conversation on board *Tulip*, which had lost most of its vitality during the long passage from Muscat. He has a slightly disconcerting but useful habit of taking nothing for granted, and soon had us reconsidering those research practices that had become entrenched through habit rather than thought.

Jonathan bends in concentration, his hunched body and cowled head suggesting a monk in deepest contemplation. But unlike the monk whose attention is directed to "the beyond," Jonathan is concerned with "the beneath" as his toes slowly twiddle the directional hydrophone. Finally he turns his head to look at the compass and calls out, "Two-ten." The whales' clicks bear approximately 210 degrees.

"How far are they?" asks Nic, who is at the helm.

"It's hard to tell. Let's try going for fifteen minutes," Jonathan replies.

As he removes the towel and headphones, Nic swings the tiller over and I ease the jib sheet, until *Tulip* moves off on a course of 210 degrees in pursuit of the whales.

Such is the routine of tracking sperm whales. If we can follow them through their daily lives and learn to interpret what we see and hear, then we may begin to glimpse their world. And in this business of tracking whales, the longer the better. Our record so far for following a particular group is twelve hours, during which time we started to discern a routine of resting, feeding and socializing, but we would dearly like to extend it to beyond a day — a week, perhaps. These two whales click continuously when underwater, blow regularly while at the surface. Their behavior closely follows the whalers' adage: "One blow for every minute under water." They seem to be moving consistently southwest and downwind, making for easy tracking by *Tulip*. We have hopes that we may be able to stay with this pair for many more hours, or, if we are lucky, perhaps days.

Tulip is sailing through the Gulf of Mannar about halfway between India and the island of Sri Lanka. At the northern end of the gulf, Sri Lanka and India are almost joined by a string of islands called Adam's Bridge. To the south, the Gulf of Mannar lies wide open to the Indian Ocean, but it is clearly bordered by the river of shipping that swings past Cape Cormorin, India, and Galle, Sri Lanka, on its way from the Red Sea and Persian Gulf to Singapore and the Far East.

In February, the Gulf of Mannar is a hot, steamy place. The weather is fickle. Strong winds, usually from the northeast, closely follow calms, and violent thunderstorms sweep through at unpredictable intervals.

During our two weeks in these waters we have found the Gulf of Mannar to be the home of many sperm whales, as well as a variety of dolphins, although they are not quite as numerous or as varied as those off the Arabian coast. Even the recurring faint songs of the humpback are present — a realm of cetaceans. The Gulf of Mannar is also to be *Tulip*'s home for the immediate future.

Five days ago, we had to decide whether to sail on to the Seychelles, as originally planned, or remain here off Sri Lanka. It was for Jonathan, whose Ph.D. was on the line, to make the decision. He carefully considered both sides:

"We really need to find the groups of females. The single animals and pairs that we have been following here in the Gulf of Mannar are interesting, but we are not going to be able to look at the most important issues until we find breeding sperms." Groups of females with their young, each containing from eight to thirty-five animals, form the focus of sperm whale social organization.

"I'd like to know if the females in those groups are related, and why they stay together," Jonathan continued. "Is there some advantage to feeding together, or do the females care for each other's calves? How tight is the mother-calf bond? And I'd really like to see the big breeding males with the groups of females. How do they interact with these groups when the time comes around for mating? And who does the mating?" There is a debate in the scientific community about the mating patterns of sperm whales. Is it the largest males who do the mating, as many scientists think, or, as one dissident has proposed, just the small, younger males? Do several males act in consort to gain control of a group of females? And how long does a male stay with a female group: hours, months, or years?

These are questions that perplex all sperm whale scientists. We cannot hope to answer all, or perhaps even any, of them during our voyages on board *Tulip*. However, if we can find the ways in which they may be answered, then our work will have been a success.

"Hopefully we'll find the big groups and males off the Seychelles," said Jonathan, uncharacteristically optimistic.

"But they hardly saw any sperms near the Seychelles during the aerial surveys that Sidney Holt was involved with, and those they did see were miles from port," I countered.[2]

"And even if we went right now, by the time we got there we would only have a little more than a week around the Seychelles before the end of this season," added Gay. "I'm getting tired of rushing from one port to the next. We never get a chance to have a reasonable amount of time in any of the wonderful countries we've visited. We spend almost all our time cooped up in ports, dealing with officials and obtaining supplies so that we can leave as soon as possible. It's so superficial, and I feel we are rude to the people who make so much of an effort to help us."

Like Gay I longed for greater depth to our traveling and

research. "It's the same with the whales. We keep interrupting our time with them so we can make the next port without falling too far behind schedule. I'd like to be able to stay with some whales, learn to study them, get to know them and see how they relate to their environment."

Jonathan was open to our arguments. "The Gulf of Mannar has lots of sperm whales and they're not too far from land, but I'd really like to see the big groups. It would also be nice to find somewhere a little cooler and more consistently calm."

Nic's impartial common sense was a great plus in helping Jonathan make his decision, as Gay and I were so clearly in favor of remaining in the Sri Lankan area. Jonathan pondered, wavered, and finally, with considerable misgiving, announced, "Let's stay here, then. It's probably better than the Seychelles, and there seem to be more whales than off Oman . . . I hope."

"These clicks sound different," says Lex, whose turn it is to perform the directional hydrophone routine. We each take a listen. Yes, there is a difference. The clicks are blurred together, and there are many more of them. Instead of the slow trot, characteristic of the two whales we had been following all day, we can hear a whole Grand National field passing the hydrophone. We sail on in the direction of the greatest click intensity, keeping a good lookout.

A few minutes later: "Blow! Blow! Blow! . . . They're everywhere!" The ocean ahead seems blotched with the transient misty bushes of sperm whale blows. Soon we can see the dark shapes of their bodies.

"Oh, look," yells Gay, "there are little ones!" Sixteen feet long makes a large animal, but a baby sperm whale. Whales, large and small, can be seen all around.

How should we watch? A scientist's dilemma: whether to examine overall structure, or the nature of its elements? Jonathan takes the individual whales, filming them on the video camera, photographing them, while I try to assimilate the form of the group. But we are faced with another scientist's dilemma:

A sick whale hangs listlessly in the water. (B. Coleman: P. Gilligan)

Phil Gilligan and Flip Nicklin were able to swim around the sick whale and take this underwater photograph of her slack flukes. (B. Coleman: P. Gilligan)

Saddleback dolphins off Arabia. (WWF: A. Alling)

A sperm whale at the surface looking and acting like a log, except for the low bushy "blow" slanted forward and to the left. (V. Papastavrou)

The log-like back suddenly flexes and arches as a sperm whale dives. (B. Coleman: J. Gordon)

Tulip *with breaching and lobtailing sperm whales.* (Dieter and Mary Plage)

The breach! (T. Arnbom)

Bryde's whale. This is our best photograph of these elusive whales. (A. Alling)

Striped dolphin leaps past Tulip off Sri Lanka. (A. Alling)

Gay Alling spent much of the time that Tulip was in port searching Sri Lanka's coastline for dead dolphins. (J. Gordon)

The huge blow and back of a blue whale off Trincomalee. The prominent ribs suggest that this whale is poorly nourished. (F. Nicklin)

The broad flukes of a blue whale are lifted into the air as the huge animal dives. (A. Alling)

The directional hydrophone from below. The hydrophone element turns horizontally inside the bomb-shaped housing. (B. Coleman: L Weilgart)

Following sperm whales. Hal listens to the clicks of sperm whales, turning the wheel of the directional hydrophone, and steers with his foot. We always look our best at sea! (A. Alling) emotional involvement with the subjects. A scientist should remain cool and objective, but care sufficiently about the work to perform it with precision and vigor. This is a difficult tight-rope at any time. It is impossible for us to treat this interacting eruption of huge, intelligent animals as though they were bacteria on a microscope slide.

There seems to be a core cluster of ten to twenty animals, closely packed. Some are small, less than a year old; others are about thirty-three feet long, the size of an adult female. Scattered around this focus at ranges of a few hundred feet are peripheral single whales, pairs and trios.

We have a momentary glimpse into the hub of the sperm whales' social life, a group of mature females with their young. These groups — or "pods," as the whalers called them — are considered by some authorities to be "harems," for there is often a large, mature male in attendance. Mature male, or bull, sperm whales are distinctive. They are about fifty-two feet long, compared with the thirty-three-foot females, and they weigh three times as much as the females. However, groups of females are not always accompanied by a lordly male, as is the case today, and occasionally there are two or more males. Experiments in which females in these groups were marked with "Discovery Tags" (stainless steel cylinders shot into the whale) and later killed, have shown that at least some females stay together over several years.[3]

The young sperm whale is born into its mother's group after a gestation period of about fifteen months. The calf suckles for approximately two years, and stays with its maternal group for another two years or so before the young males, and possibly the young females, as well, leave to form more temporary "bachelor" groups. The females become sexually mature at roughly nine years of age, but the males, who are busy building up their enormous bulk, do not reach sexual maturity until about twenty-six. Sperm whales may live to about sixty years, although heavy whaling probably means that there are few such veterans swimming today's seas.

The groups of females with their young spend the entire year in tropical or subtropical waters. In contrast, the males, as they grow, move to colder and colder oceans, so that very large males are found close to the polar pack ice. However, they presumably return to the tropics for a few months each year to mate.

Such is the skeletal view of the sperm whales' lives that the whalers have left us. Our task on board *Tulip* is to investigate how benign research might complete the picture. Can we describe the daily life of a sperm whale, trace the social interactions within the large groups, or watch males taking possession of their "harems"? The I.W.C. scientists who examine sperm whale populations are gravely hampered by a lack of knowledge about sperm whale social behavior, or the "Sperm Whale Model," as they call it. If they are to predict with reasonable accuracy the response of sperm whale populations to past, present (sperm whales are hunted today off Japan and in other parts of the world) and future whaling, or to protection, then they need a basic knowledge of sperm whale social structure.

Lack of this very knowledge has already proved jeopardous. Much recent sperm whaling has concentrated on mature males. Harpooners naturally selected the larger, more valuable animals. They were encouraged in this selection by I.W.C. scientists who subscribed to the "harem" view of sperm whale social organization, and assumed there must be "surplus males," who could be removed without harming the reproductive rate of the population. However, in the late 1970s evidence began to appear that in some areas fewer female sperm whales were becoming pregnant as the males were killed off. Something is wrong with the Sperm Whale Model, but no one knows what.

One of the justifications of the *Tulip* Project is to try to fill in the gaps that whaling science could never reach. Today's observation is a small, but encouraging, start.

The whales are beginning to behave peculiarly. Two smallish (twenty-eight-foot) whales slowly but simultaneously raise their smooth black bulbous foreheads above the water, about

THE CASE

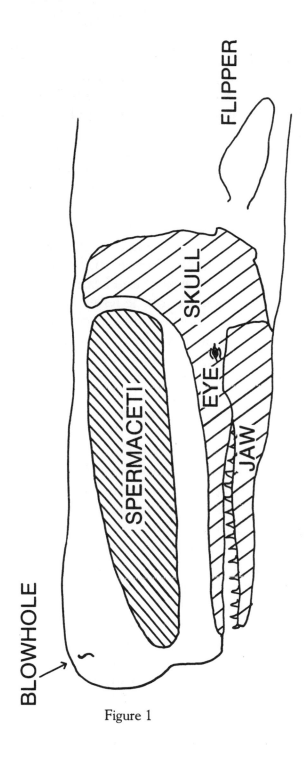

Figure 1

four yards apart, and then remain "for some time in that position, solemnly bobbing up and down amid the glittering wavelets like movable boulders of black rock," as Frank Bullen described the curious behavior of "spyhopping" in *The Cruise of the Cachalot.*[4]

What do these spyhops signify? Perhaps they allow the whales, whose eyes are situated on small promontories on each side of the head, to watch one another with binocular vision? And what is the significance of that strange forehead — one of the most unusual organs in the animal kingdom? The whalers called it the "case," and it forms a vast reservoir of spermaceti, the finest grade of oil known to man. It is positioned above the jaws and in front of the skull and brain. In a large male it may reach almost one-quarter of the length of the whale.

A mistaken view of the function of the oil gave the sperm whale its name, while its market value drove the whalers around the world, and the sperm whale toward extinction. But what is this "case"? (fig. 1) Melville considered it to be a battering ram,[5] and he may well be right, although it is unlikely that this would be the sole function of such a complex and unusual organ. Modern scientists are divided in their opinions. A distinguished American whale scientist, Dr. Kenneth Norris, believes that it allows the sperm whale to modify or focus its clicks.[6] The clicks are probably so important in the lives of sperm whales that even such a strange appendage as the spermaceti organ might be favored in evolution if it increases their efficiency. The principal rival theory is espoused by Dr. Malcolm Clarke of England, who suggests that the sperm whale can control the temperature of the spermaceti, allowing it to solidify or melt.[7] This would change the whale's buoyancy, and permit it to ascend or descend with little effort. These two ingenious theories are hotly debated. But for now, neither has ascendency and the whale retains its mystery.

Behind the case lies another mysterious organ — the sperm whale's brain, the largest of any animal. We have no clear idea what it is used for. The study of brains is difficult, even with small captive animals. The brain of the sperm whale, larger than our own and attached to a huge, mobile animal, is for the moment beyond reach of all but postmortem study. However, it seems reasonable to speculate that a fair proportion of it is engaged in processing sounds heard from the ocean, including, almost certainly, the echoes of its own clicks and the clicks of other sperm whales.

The short tropical twilight has arrived, and dark will soon close this encounter. The shining twilight ocean is suddenly ruptured as a sperm whale bursts through the surface and into the air. Briefly the huge animal hangs above us before crashing back into the ocean. This "breach" is the most powerful display of any animal. It is admirably described by the early nineteenth century whaleship surgeon Thomas Beale:

> One of the most curious and surprising of the actions of the sperm whale is that of leaping completely out of the water, or of 'breaching', as it is called by the whalers. The way in which he performs this extraordinary motion, appears to be by descending to a certain depth below the surface, and then making some powerful strokes with his tail, which are frequently and rapidly repeated, and thus convey a great degree of velocity to his body before it reaches the surface, when he completely darts out.[8]

Beale's admirer, Herman Melville, embellished this description in his account of "the wondrous phenomenon of breaching":

> Rising with his utmost velocity from the furthest depths, the Sperm Whale thus booms his entire bulk into the pure element of air, and piling up a mountain of dazzling foam, shows his place to the distance of seven miles and more. In those moments, the torn, enraged waves he shakes off, seem his mane; in some cases, this breaching is his act of defiance.[9]

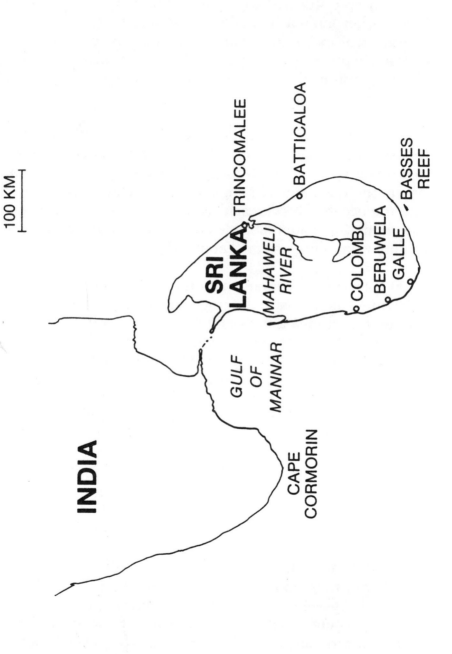

100 KM

INDIA

SRI LANKA

TRINCOMALEE

BATTICALOA

MAHAWELI RIVER

COLOMBO

BERUWELA

GALLE

BASSES REEF

GULF OF MANNAR

CAPE CORMORIN

5

GULF OF MANNAR, SRI LANKA

February 28 - March 15

"DOLPHINS!" RINGS OUT FROM ON DECK.

I'm in *Tulip*'s galley, trying to prepare the evening meal. I brace the various pots and pans in case *Tulip* lurches suddenly, unhook myself from the strap that holds the cook in the galley area (leaving hands free for food preparation), pick up a small and wizened Sri Lankan eggplant that had fallen from one of the swaying hammocks that hold our vegetables, and start a search for a vagrant tin of tomato paste that has been rolling noisily around the cabin floor. But Gay, who is making her way down the companionway for her clipboard and camera, says, "There's no hurry, Hal. I've only seen splashes on the horizon."

However, the dolphins must be moving fast: by the time the ingredients and cooking utensils are secured and I reach the deck, we can pick out individual forms leaping amid the spray. Then suddenly, whole ranks of hundreds of animals come streaming from the water in an extraordinary display of precision movement. This "porpoising" behavior, in which the dolphins make long horizontal leaps over the surface, has been calculated to be a particularly efficient method of traveling fast when an animal needs to stay near the surface, as dolphins must to breathe.

One animal leaps vertically, clearing the water surface by several body lengths time after time — a repeated thrust at the heavens.

"Striped, aren't they?" asks Nic, ready with her camera in case they come closer.

"They sound like it," replies Gay, who has briefly monitored the hydrophone.

Over the weeks we have grown to know the dolphins of the Gulf of Mannar. This exceptionally dramatic aerial behavior, and the babel of squeals on the hydrophone, tell us with reasonable certainty that these are *Stenella coeruleoalba*, the lovely striped dolphin. And sure enough, as they approach, we see the characteristic delicate, dark waterline stripes, and the subtle blazes sweeping up and over the backs of the ten-foot-long dolphins. Many of these animals also have bright pink bellies. Rushing over the waves, bound for some unknown goal, the dolphins are out of sight within a few minutes, although the attendant squealing on the hydrophone lingers a little longer.

The striped dolphins are the most offshore of the species we regularly sight in the Gulf of Mannar. Nearer the continental shelf, and occasionally over it, are the slightly smaller spinner dolphins, *Stenella longirostris*, whose variety of aerial behavior so delighted us off Oman. The spinners are also quite common here. We have made some attempts to follow them, and have observed the different phases of their daily life: lolling at the surface, scurrying around busily in small groups of about ten animals — perhaps feeding — and sweeping the ocean in large coordinated, fast-moving groups of several hundred. It is during this last phase of activity that we generally see the spins, leaps and cartwheels. The spinners make slightly higher-pitched squeals than the striped dolphins; they also have a "chatter," which sounds rather like a distant human cocktail party. Being somewhat slower, they are simpler to track than the striped dolphins, and Gay and Jonathan have had the good fortune to swim with them briefly.

Along the edge of the continental shelf, as off Oman, we see the larger gray Risso's dolphins, *Grampus griseus*. Their vocal output contains mainly clicks and "pulsed tones" — clicks repeated so fast they sound like squawks — with few of the pure-toned squeals so characteristic of their smaller cousins. These clicks are shorter and at a higher frequency than the sperm whales'; they are often heard in bursts, which is frequently a characteristic of echo-location.

And on the shelf, in the shallow water closest to land we have seen small (about six feet long) gray dolphins. They seem to be a relative of the familiar bottlenose dolphin, *Tursiops truncatus*, but are much smaller, less vocal and less demonstrative. We see no leaps or twists or tail slaps from these tranquil animals. They may be the controversial Indian Ocean bottlenose dolphin, *Tursiops aduncus*, which some scientists do not believe exists.

During our port calls in Sri Lanka, Gay has several times traveled to Beruwela, a fishing village forty miles south of Colombo, to investigate dolphin mortality in fishing nets. Here she has seen the small *Tursiops*. They had become entangled in fishing nets and had died gruesome deaths. The fishermen set large drift nets at night from their thirty-foot-long boats to catch tuna, sharks and other large fish. But these nets also catch turtles, sea snakes and dolphins. Although there is some directed hunting for dolphins and we have seen light harpoons being thrown unsuccessfully toward a Risso's, most fishermen do not wish to catch them. They receive little money for dolphin meat, and some regard it as bad karma to kill such developed animals.

Sri Lanka, although principally Buddhist, has sizable populations of each of the world's four major religions. The Muslim and Christian fishermen seem less loath to catch dolphins than the Buddhists and Hindus. We were told of a Hindu legend about a powerful king who threw a ring into the ocean, telling his followers, "He who brings back the ring will have the hand of my daughter." The knights who dived in became dolphins, whose narrow snouts are shaped to pick up the ring. I do not recall whether any succeeded; perhaps they found life as dolphins much more rewarding than anything they could imagine as the son-in-law of a king.

But the dolphins continue to die, perhaps over one thousand a year in Beruwela alone. During the rest of the *Tulip* Project, Gay hopes to be able to document this by-catch, to calculate approximately how many are being caught and of which species. This may allow her to estimate the possible effects on the dolphin populations and to see what might be done to alleviate the by-catch.

Dolphin by-catch is a worldwide problem, and as drift-net fishing with strong artificial fiber nets increases in many developing countries, the scope of the problem is almost certainly expanding. Dolphins, who had passed straight through the traditional natural fiber nets, leaving a rent that could be easily repaired, become inextricably entangled in the petroleum-based modern counterpart. The destruction may be so extreme that some stocks or species have become endangered, but we know virtually nothing. Perhaps the by-catch in Sri Lanka can form a microcosm where Gay can gain some insight into this unintentional but gruesome slaughter.

During the past three weeks we have encountered our share of disappointments and troubles. Roger Payne was to join us. We were looking forward to seeing what his remarkable insight on whales could add to the study. And he was to bring with him a more sophisticated directional hydrophone, which we hoped would allow us to find and follow the whales much more efficiently. But Roger ran into insurmountable problems, both personal and with his work. So after Lex departed, there were only four crew.

It proved hard to follow the whales for long periods with our primitive directional hydrophone and only four crew members, all of whom were worn out. However, we did manage one watch of a group of sperm whales lasting over twenty-four hours, during which time the whales, and *Tulip*, traveled over sixty miles. We left the sperms to return to Colombo in the hope that Roger might have broken the shackles of land. No luck.

The sperm whales we followed were mostly in groups of one to three, and they meandered around the Gulf of Mannar, with no obvious goal or direction. They usually moved slowly, at less than two knots, so that *Tulip* was never troubled to keep up.

The whales seemed to have two lives: the one we saw when they were at the surface, and another, shrouded in mystery, a quarter of a mile beneath. Mostly their times at the surface seemed purely to satisfy their mammalian need to breathe air. Usually they would come up, lie almost dormant at the surface blowing regularly and, when ready, descend to their business beneath. However, sometimes they would spend more than the seemingly requisite five to fifteen minutes at the surface — perhaps an hour or two. And then we could glimpse their social behavior, as individuals edged close to one another, rolled, spyhopped or breached. But sooner or later they would dive again.

Once they were in the depths, their behavior was even harder to grasp. Sometimes on our recording depth sounder we would see a dark, thick arch, representing a whale underwater. The depth sounder, true to its name, would register the whale's depth, usually about fifteen hundred feet directly beneath *Tulip*. While at the surface the whales were usually silent, whereas when they are deep we would hear the steady clicking, which would sometimes show as dark flecks on the depth sounder record.

What are these clicks? The most plausible explanation seems to be that they are used for "echo-location." Like our depth sounder, the whale emits bursts of sound, then waits for an echo. The sperm whale probably views the ocean environment principally by using the sounds it hears, especially the echoes of its own clicks, to form an acoustic "picture" of its surroundings. Unlike light, whose images are mainly produced by reflection and scattering from the surfaces of objects, emphasizing the external, sound penetrates. Thus, in common with doctors who use ultrasound to examine internal organs, sperm whales may well be able to inspect one another's insides (and perhaps boats' or human swimmers' insides, as well) from the echoes of their clicks. Thus they may know of their companions' feeding success, diseases, stress or fear. Little can be hidden in a society where all can know the interior of each. Could it be that sperm whales live in a society where deception is impossible?

But what is the sperm whale principally "looking" for with its clicks? It could be the sea bottom, the surface, other sperm whales, or even *Tulip*. We gather clues. The whales do not seem to click much at the surface. During the half second between clicks, a sound travels about twenty-six hundred feet in seawater, which suggests that the whale is interested in things less than thirteen hundred feet away, so that the echo of one click will have arrived before the next one is made. And the whales click regularly and almost continuously, which suggests they are trying to keep constant track of something. What?

For me, the most likely answer seems to be food. As they dive, the whales are scouring the nearby ocean with their clicks for the squid on which they depend. It has been calculated that an adult female sperm whale would have to catch about three hundred squid a day, each of about three pounds, to fulfill the requirements of her huge body. For an adult male these requirements are tripled.

So the depths are for feeding, and the surface for breathing, with a little socializing thrown in. As I ponder this dual life-style and try to picture that wrinkled body surging through the deep dark ocean, clicking for squid, an idea emerges: maybe the whales look different when they are far below the surface. The U.S. Navy trained a bottlenose dolphin called "Tuffy" to dive to considerable depth and take its own picture by pushing a button. The photographs show Tuffy's normally lithe, stream-lined body flattened and distorted by the huge pressures.[1] Perhaps the opposite happens with the sperm whale, so that the pressures of the deep flatten the ridges and bulges, and the wrinkled "ugly duckling" of the surface is transformed into a graceful "swan" of the abyss.

When we left Colombo for our final spell at sea on March 13, we restricted our ambitions to sailing a transect over the Gulf of Mannar, in an attempt to document its teeming cetacean life. We drew a simple grid pattern over a chart of the gulf with "stations" every ten nautical miles, at which we would listen through our hydrophones for the sounds of whales and dolphins.

But it was an ill-fated transect. There were plenty of whales and dolphins to hear, and a few to see, but after one day our difficulties started: both tape recorders developed problems, and then the wind rose from the northeast, making most legs of our grid a beat upwind. As *Tulip* slammed into the seas, Jonathan and I tried to adjust the delicate innards of the tape recorders.

Then the engine stopped. I fixed it — I forget how.

We had made our way across the gulf to the shallow coastal shelf off India. In the steep swell of the coastal banks we all felt ill. Then the engine stopped again, with the mainsail splitting almost immediately afterward. While Jonathan, sick with flu, sewed the sail back together, I attacked the engine. After eight hours of work, during which I replaced both cylinder heads, it finally ran. We continued the transect, but the mainsail repairs still had far to go.

We turned a corner of our grid, back east toward Sri Lanka, but the wind headed us, and soon the engine had lost all oil pressure. I began to dismantle it once again, but then Nic announced,

"Well, everybody, I believe we're out of tea!"

I looked up from the engine. "It's too much. The Fates are

definitely against this transect. Let's head for Colombo."

If the wind holds, we will be in Colombo late tonight and will have finished this season's work at sea. The gradually worsening weather signals the beginning of the southwest monsoon season and the end of this phase of the study. There will remain a few days of writing our interim report, cleaning *Tulip* and arranging for her to be laid up. Then we shall leave Sri Lanka to return to the North Atlantic by air: Jonathan and Nic to England and analysis; Gay to the dolphins along the cold, rocky shores of Labrador; and I to the humpback whales on the fog-filled Grand Banks of Newfoundland.

While waiting for the dinner to cook, I sit on the pulpit at *Tulip*'s bow and watch the misty sunset over the ocean. During the months since we left Greece the pure adventure of this voyage has many times been eclipsed by struggles with the obstacles we found in our path, by frustration, heat and exhaustion. My goals evolved from seeking the unusual, to pursuit of understanding. We have played out the attraction of exotic lands and new species; the fascination now is to understand the sperm whale.

I contemplate the achievements of *Tulip*'s first season. In finding a research area where there seem to be many sperm whales, as well as dolphins and humpbacks, we have achieved the major objective of this preliminary voyage. It is hot and sometimes rough here in the Gulf of Mannar. But because of Sri Lanka's narrow continental shelf, the sperm whales are comparatively close to shore. Colombo Harbor has many of the attributes of a cesspool, but perhaps we can find another smaller and more pleasant port from which to work. Sri Lanka seems a reasonable base: it is relaxed, cheap, and despite the poverty, there is a consciousness of conservation and a high level of education that would be the envy of many more developed countries.

We have found that we can follow sperm whales for periods of over a day and can identify some of them individually from marks and scars on the dorsal fin and flukes. It would be of considerable significance if we could do both more efficiently. A better directional hydrophone and experimentation with cameras, lenses and film types might help. Perhaps we could also devise schemes to measure the whales at sea, or to identify them acoustically from their clicks. Jonathan and I will work on these technicalities during the eight months before returning to *Tulip* in January 1983. And if we could only find more of the large groups of female sperm whales . . .

It is clear that we can learn some things of the sperm whale using our benign techniques, but how much? Will we ever be able to gain an insight into the lives of these enigmatic animals comparable to that which has come from the years of research on southern right whales and humpbacks? There is certainly promise, but also much uncertainty.

Part II

Off Sri Lanka
Winter 1983

6

GULF OF MANNAR, SRI LANKA

January 5 - 22

I JUST WANT TO TUMBLE INTO THE WOMB of sleep. No, Hal, you have work to do. Not difficult — just stay awake and watch the satellite navigator, which magically transforms signals from unseen satellites into positions on the ocean. *Tulip* lurches, and I clutch the raised fiddle (ledge) around the chart table. I long for peace, but can find none. We arrived in Colombo two and a half weeks ago for *Tulip's* second season with the whales. The tropical warmth, palm trees and friendly Sri Lankans who greeted us in Colombo were especially welcome when contrasted with icy (in more ways than just temperature) Moscow airport in January (Aeroflot is the cheapest way to reach Sri Lanka from Europe). But very soon I felt stifled by the heat and bureaucracy, oppressed by the ever-present harsh sounds of the city and by the endless bargaining, frustration and delays. And even here, in the sea, which usually brings tranquillity, we simply endure. *Tulip* struggles over the waves. Through the hydrophones we occasionally hear dolphin squeals or a series of sperm whale clicks, but I know we need to be achieving more.

    Tulip has the wind on the beam, which is a favorable point of sailing. A new working jib, replacing "Yves's sail," straining at its sheets, leads the tiny fully reefed mainsail in powering us over the waves. In front of me the needle of the knotmeter (nautical speedometer) reaches up to seven knots for a few seconds as we surf down a wave, then settles back to just over five. At least at this rate the transect, a rerunning of the aborted survey of the whales and dolphins of the Gulf of Mannar that we tried at the end of last season, will soon be over. But then what? It is much too rough for the detailed behavioral research that is our major task. We will just have to wait for the weather to improve; at least we know there are sperm whales here.

To my left a two-tiered digital display on the satellite navigator glows red. The upper part reads 8°20.0'N — our latitude. Beneath it the longitude flicks from 79°8.3'E to 79°8.4'E. I twist around, grasp two handles on either side of the companionway and haul myself out.

"Another one point six of a mile to the station," I tell our watch keeper, Margo Rice, one of the two new crew members who have joined Jonathan, Gay and I this season. Margo is sitting on the windward (upper) cockpit coaming, a tall woman in her early twenties, dressed in shorts and a yellow oilskin jacket, with the hood raised so that it covers most of her light brown hair.

"Should take us about fifteen minutes," I add. "How are you doing out here?"

"Oh, pretty good." But Margo's usually beaming smile is tinged with weariness.

"I bet you wish you were back in Hawaii."

"Oh, no!" replies Margo. "Well, at least not yet. It's funny, but when we were staying in a luxury apartment, studying whales from a clifftop in perfect weather, we all used to long to be bouncing around in a small uncomfortable boat far from civilization but close to the whales — 'real whale research,' we called it." Margo is referring to her experience two years ago on the large and successful study of the humpback whales off Hawaii run by Jim Darling and Peter Tyack. Margo and I, being first cousins, have known each other since childhood. She joined *Elendil*, which has become *Tulip*, for the study of the humpbacks in the West Indies that we made during the winter of 1980. She worked with us again in Newfoundland the following summer, and went on to take part in whale research projects with others in Bermuda, the Gulf of Maine, off eastern Canada, Hawaii and, most recently, southeastern Alaska. Through all of this she has been a prized companion because of her gentleness, quick intelligence, warmth and cheerful quiet humor.

I peer below at the satellite navigator, "Nine point seven — almost there." When the longitude reads 79°10.0'E we will be at the next station.

"I'll do the water temperature," says Margo. She picks up the plastic bucket tied to a long rope, which has been rattling around in the cockpit. Leaning against the rail, she carefully throws the bucket forward, mouth toward the water. As the bucket passes beneath her, she hauls it up quickly, half full, over the rail and into the cockpit. The bucket must be hauled with some care, especially when traveling at speed, as *Tulip* is now. If it fills with water and is not hauled aboard immediately, it acts as a sea anchor and may break the rope, the bucket, or even pull the crew member overboard. But Margo has had much practice in filling buckets at sea. I drop a thermometer into the water that she has hauled and check the satellite navigator again.

"Ten point zero — right on. Shall I do the down-below stuff?" I ask.

"If you wouldn't mind. I'm okay up here, but . . ." Working down below is a sure way to bring on seasickness for those who are susceptible.

As I clamber down to the stuffy cabin, Margo disconnects *Tulip*'s tiller from the self-steering gear, Clarence, and thrusts it hard over to leeward. *Tulip* comes up "into the wind," then falls back — "hove to." Down below, I turn on the acoustic system and listen. Mainly I hear the splashing of waves and the surging of the hydrophone, but high above the pandemonium soar the whistles of distant dolphins. After I have listened for three minutes, Margo returns *Tulip* to a course for the next station. On the data sheet, securely anchored to a clipboard, I tick the column marked Dolphins? but place zeros under Sperms? and Humpbacks? I hand the clipboard up through the companion-way for Margo to fill in the water temperature and other environmental information. I feel too drained to follow, although I know the climate and motion would be much less oppressive on deck.

I stare listlessly at the contents of the chart table in front of me. It is the usual jumble of charts, pencils, the British Admiralty sailing directions for the east coast of India, pencils, a novel and miscellaneous electronic parts. But it is the charts that catch my eye. I love charts and can stare at them for hours, feeling the promise of unexplored lands, planning passages, or pondering the whales that may live beneath each square of paper. In front of me lie the British Admiralty routing charts for the Indian Ocean in January and February. They tell the expected weather conditions for each sea area. Different-colored arrows represent currents and winds, while contours show temperatures, wave heights, the probabilities of gales and, although not relevant to these waters, the limits of pack ice. These charts conjure up the great sweep and variety of the oceans. I can empathize with Ahab, whose charts allowed him to thread "a maze of currents and eddies, with a view to the more certain accomplishment of that monomaniac thought of his soul"[1] — the sperm whale.

On the January routing chart, a long, thick red arrow pointing from the northeast into the Gulf of Mannar represents the northeast monsoon — the winds that are giving us such a rough passage at present. The seasons in Sri Lanka are dominated by the two monsoons: the stronger, wetter, southwest monsoon, which blows from May until September, and the northeast monsoon from November to February. We have centered our research on the breaks between the monsoons, when the weather is usually calmest. The question to which I would dearly like an answer is: When will the northeast monsoon end to bring calm weather so that we can start our detailed sperm whale behavioral research? On the February routing chart the arrow from the northeast is still substantial, but it is less long (indicating fewer northeast winds) and thick (indicating lighter winds) than in January. Nothing definitive there. I suppose we shall just have to wait and see.

My eyes stray across to the east coast of Sri Lanka. The wind arrows are thinner there. If this wind keeps up, maybe we should sail around to the east coast and attempt our research there. The scientists at the Sri Lankan National Aquatic Resources Agency (N.A.R.A.) have told us that the area off Trincomalee, the principal port on the east coast, has an interesting submarine canyon close to shore. Such features often attract sperm whales.

My studies are interrupted as a bowl containing eight garlic bulbs is deposited on the routing chart. Martha Smythe, our second new crew member, is about to start cooking dinner.

"Hi, Hal-baby. Could you pass these up to me?"

Martha's South Carolina drawl is tempered from years spent in Michigan. She is a slim woman, with curly brown hair, a round friendly face and impishly pointed chin.

"Sure. How do you like sperm whale research?" I ask.

"It is wild!" replies Martha gleefully.

Unlike the rest of us, Martha seems to have no adverse reaction to the weather, so it is appropriate that it is her turn to cook — a terrible task for anyone feeling at all queasy. Three days ago, when trying to cook as we were beating to windward through a heavy sea on the first day out, I was seasick for the first time in fifteen years of sea going.

Martha Smythe has the least experience of any of us in sailing offshore, but she has spent much of her life around boats. She became a friend of Gay's during their time on the research/training schooner *Westward* (which also forms part of Margo's wide experience), and since then has taken part in two seasons of blue whale research with Richard Sears in the Gulf of Saint Lawrence. Our exhaustion at the end of last season suggested that we really needed five crew members to operate efficiently. We also felt the lack of an outgoing, effervescent personality who could help us stay cheerful. Gay assured me Martha would be able to supply this, and she appears to have been right.

With Gay and Jonathan, who are asleep in their bunks, we seem to have an experienced, well-balanced crew. Now if only the gods of wind and wave will let us begin our real work, for which we have been planning for so long.

Wheels have been turning about our work in the past few months. Two months ago, with the crew for this season selected and many of the supplies bought, W.W.F. suddenly announced they were going to reconsider our funding. We had no idea why; scientific reaction to our first season had been generally favorable. A meeting was called in New York, at which Sidney Holt revealed the root of the problem: "The officials in the Seychelles were under the impression that the *Tulip* money was raised for work in their waters, and aren't pleased that they haven't seen you." I was flabbergasted, having no conception that such expectations surrounded our project. A compromise was reached whereby we would sail *Tulip* to the Seychelles for a scientific meeting due to take place in May 1983. It would principally be a courtesy visit, but it should smooth the ruffled feathers.

Another development that worries me is "the film." W.W.F. have contracted a smooth-talking New York film producer to publicize *Tulip*'s work. The film crew is due to arrive in a month, by which time we will hopefully have found something more interesting for them to film than the tops of waves.

During the interval between our first and second seasons *Tulip* was moored in one of the freshwater lakes in central Colombo, beside a small boatyard — Banno Lanka, Ltd. There was a fair amount of work that we needed to do on her — repairs, maintenance and installing the new equipment that we had brought out. For much of this we would not need all five crew members, so Gay and Martha traveled to the north of Sri Lanka to investigate the dolphin by-catch there.

While Margo cleaned out the lockers, Jonathan screwed alloy mast steps on *Tulip*'s spar (these would allow us to climb the mast much more easily), and I installed the BBN. The BBN, named after the American company that built it (Bolt, Beranek and Newman), is a large case containing preamplifiers and an oscilloscope, which we hope will allow us to distinguish individual whales from their clicks.

Margo's program of reorganizing *Tulip*'s lockers turned up a pile of junk, which even the Sri Lankans, who manage to find value in most of our garbage, had no use for. So Jonathan made a fire on the wharf and I carried down the bags of rubbish. Even in the heat of Colombo the flickering flames of a controlled fire are a pleasant sight, but as I watched the blaze Jonathan cried from *Tulip*, "Where are the sail-making supplies?"

"What were they in?" I asked.

"A brown paper bag."

"Oh, no!" I looked into the fire, and remembered some strange green flames — sail patches? We doused the blaze quickly, and sure enough, in the ashes were distorted sail needles and a charred piece of leather that had been a sailmaker's palm. What else had we lost?

We were given considerable assistance in the refitting of *Tulip* by the workers at the boatyard. They were Tamils from the northern part of Sri Lanka. The Tamils, who are Hindus, are a distinct race from the Buddhist Sinhalese who predominate throughout most of the island. The two peoples also have different languages and scripts and have never been particularly friendly toward one another. Banno Lanka, like the British rulers a hundred years earlier, value the Tamils for their high level of education and reputation for hard work. Many Tamils have been brought into the Sinhalese parts of the country to work as skilled and unskilled laborers. This has fanned the resentment of the Sinhalese, and the mutual distrust of the two races.

In the evening, after our work, we would walk to the Krishna Palace, a tiny Tamil vegetarian restaurant where the three of us would eat the spicy food with our hands, in the South Asian manner. The bill for three full meals, with split-second friendly service and enough bananas for breakfast, was usually a little over a dollar.

At night I would lie on deck. I could watch bats feeding on insects above the lake, or listen to a different predator — a Sri Lankan fishermen paddling his nightly rounds in a dugout canoe. Sometimes the shouts of the laborers working in a neighboring warehouse would keep us awake, but mostly we slept soundly after our hard work.

Tulip needed to be hauled from the water so that her bottom, encrusted with a forest of marine and freshwater growth, could be cleaned and painted. It proved difficult to arrange for this, but finally the boatyard found a mobile crane, which one morning made its way along the wharf, put out its stabilizing legs and yanked *Tulip* out of the water. The Banno Lanka workers scrubbed and painted industriously, so that by late afternoon she was ready to return to the lake. The crane picked *Tulip* up and began to swing her out. I was on board so that when she had been returned to the water I could release the straps and motor her round to the dock. About halfway out *Tulip* began to drop, fast; the crane was toppling over under the weight of the boat. The additional coat of paint together with my presence must have somehow tipped the scales. At that point we were suspended over a mud bank that separated the wharf from deep water. *Tulip* came down, splat, with the crane canted over above her. What a way to start a study of sperm whales — sitting on a mud bank in a freshwater lake in central Colombo! The Sri Lankans thought the situation funny, but were unperturbed. This kind of incident was part of their daily life: just beside *Tulip* lay the ruins of a barge that had failed to make the water on its maiden launch, and had lain half in and half out of the lake ever since. After some maneuvering, the crane repositioned itself and within an hour had managed to complete *Tulip*'s launch.

A few days later *Tulip* was once again stuck in the mud. We had finally obtained permission to take her back to to the harbor, bribed the keeper of the lock that links the freshwater lake to a seawater canal and were making our way out through that canal. However, the canal, which runs directly through central Colombo, had silted up with mud, sewage and all the rubbish that had been thrown into it over the years. *Tulip* plowed laboriously through the quagmire, and finally came to a halt underneath a bridge carrying one of Colombo's principal thoroughfares. We raced the engine at full throttle, but the ship did not move. Soon the engine overheated, as the cooling-water intake clogged with mud. The traffic thundered overhead; the murky waters gurgled beneath. We sat and waited for the engine to cool and the tide to rise. Oh, whale research!

Somehow we made it through and reached Colombo Harbor. *Tulip* then had another tussle with a crane. This one was trying to raise her mast. But we were nearly ready for the sea.

Gay and Martha returned from the north bearing large plastic bags full of something heavy. There was a rank smell about them, discernible even above Colombo's usual powerful blend of odors.

"What have you got there?" we asked.

"Dolphin skulls. We were popular on the train! There's a fairly big by-catch in Trincomalee, and we were able to examine a few animals. By the way, Trincomalee is lovely, and the fishermen sometimes see whales," they added with large grins, clearly hinting they would not be averse to *Tulip*'s course being directed that way.

The skulls were buried in the grounds of Banno Lanka while

we waited for arrangements to be made to export them to the U.S.A., where they could be examined by the experts most qualified to review the taxonomic status of Sri Lankan dolphins.

The sedate world of taxonomy seems so far away as *Tulip* heaves up on a surging wave. In the galley, just across the boat from the chart table, Martha's cooking is producing fine smells (garlic dominating). Soon it will be dinner — the most social time of day on *Tulip*, and usually eagerly anticipated. But tonight I suspect our appetites will not be so good.

7

COLOMBO TO TRINCOMALEE, SRI LANKA

January 23 - February 17

TULIP MOTORS SOUTH OVER SMOOTH SEAS, her engine rumbling steadily. As the sun climbs, it beats fiercely on the deck, and the heat builds up here below. But in this benign weather we can leave the hatches open, and light Terylene (Dacron) "wind scoops" are rigged to channel the gentle breeze caused by *Tulip's* motion down below. The grace of calm.

The regular hum of the engine is broken as Jonathan, who has the watch on deck, throttles back and pulls the stop cable. The engine splutters to a halt. It is time to listen. *Tulip* slows, and the wind scoops collapse. Almost immediately the atmosphere in the cabin grows oppressively sultry.

Jonathan pays out the hydrophone, then swings easily down to the chart table, where he flicks a few switches and dons the headphones. He listens intently for a minute. I watch. Removing the headphones, he smiles at me. "Nothing." Neither of us want to hear the whales — not yet. During the past three weeks we have been overwhelmed and exhausted with good fortune.

There have been no outstandingly dramatic events, but we have lived, and lived fervently, with the whales. This peaceful pause is more than welcome.

During our survey of the Gulf of Mannar in January we found some sperm whales, but the weather was too rough for behavioral research. So, after putting in to Colombo for supplies, we headed around the southern end of Sri Lanka toward the east coast and Trincomalee, where we hoped to find calmer weather and cooperative whales.

We have been blessed beyond all such hopes. After passing the Basses reefs at the southeastern corner of Sri Lanka, famed for their rough seas, the strong winds disappeared. Since then, there have been either calms or gentle breezes, allowing *Tulip* to sail easily or motor with the sperm whales, which have been plentiful off almost the whole of the east coast.

Despite the intensity of our work and consequent lack of relaxation or sleep, the crew have kept up their spirits, and our different personalities mesh well. Martha's exuberance sets the tone. Gay, who is close to Martha, seems almost released from care. She is happier and less serious than I have ever known her, and we often hear a previously unfamiliar boisterous laugh ringing through the boat:

"Oh, Hal, it's wonderful! Frankly, I thought the sperm whales were rather dull last year, and was not particularly looking forward to this season. But they are fascinating animals!"

Jonathan, realizing that he has the means to make a fine study and excellent Ph.D., has become a creative dynamo, constantly trying to devise and experiment with new techniques of studying the whales. My own scientific inclinations, which are strongly based on the rigorous collection of standard data, act as a stabilizing influence. Complementing each other, we experiment with Jonathan's ingenious approaches, but try to keep our methods as systematic as possible. We both feel the thrill of scientific breakthrough, and enjoy the creative interaction with each other's mind.

"Blow!" shouts Jonathan.

I hurry on deck. However many spouts we have seen in the past three weeks, I never tire of the enigma of the whale.

"It seemed a straight blow," says Jonathan. "Probably a Bryde's. There it is!"

A vertical pillar of spray rises about eight feet from the water two hundred yards off the starboard beam. The blow is considerably more concentrated and powerful than the rather languorous exhalations of the sperm whale.

About thirty seconds later there is another blow, and this time we see the dark gray animal as it scythes through the calm ocean. A V-shaped head topped with three "rostral" ridges precede the blowhole and identify the animal as a Bryde's whale (*Balaenoptera edeni*). Then follows a smooth, slim back, topped by a moderate-size, sickle-shaped dorsal fin with a ragged trailing edge. Jonathan quickly photographs the distinctive fin; perhaps we can compile a directory of individually recognizable Bryde's whales, as a companion to the sperm whale catalog. The whale is not sighted for another three minutes, but then we see a blow eight hundred yards off the stern. The same animal? These whales are elusive, and I wish luck to anyone trying to track them.

The Bryde's whale is perhaps the least known of the large whales. The existence of the species was only confirmed in 1913 by the Norwegian scientist Professor Ø. Olsen, who was working in South Africa.[1] He named the whale after Johann Bryde, the Norwegian consul to South Africa, who had assisted his research.

The streamlined body and graceful movements contrast with the jerky slothfulness of the similar-size (about thirty-three feet long) female sperm whales. Unlike the sperm whales, which are built for the depths, the Bryde's whales seem to stay near the surface. Here they chase and catch their prey, principally schools of small fish, in single gulps. The water that is taken in inadvertently with the fish school is strained out through the whale's baleen, two rows of fibrous plates hanging from the upper jaw. Most of the large whale species, including the right, gray and humpback whales, have this baleen. They are the Mysticeti, Greek for "mustached whales." In contrast the Odontoceti, which include dolphins and porpoises and whose largest member is the sperm whale, have teeth.

Of all the baleen whales, the Bryde's whale is the most tropical, and so it was the one we most expected to see in these waters. During our three weeks off the east coast of Sri Lanka, we have seen them almost every day, cruising erratically in ones and twos in the productive waters near the edge of the continental shelf.

But the Bryde's is not the only baleen whale of these waters.

On February 7 we made port in Trincomalee for supplies, repairs and, of course, bureaucracy. We put in late at night, and not being sure of the intricacies and regulations of the large and complex harbor, we anchored in one of the coves that fringe it. In the morning we awoke to find ourselves surrounded by clear, clean water leading to steep shores covered with luxuriant jungle. Birds called and monkeys swung from the trees. Paradise? It certainly seemed so when compared to the urban filth and noise of Colombo.

Further acquaintance with the beautiful harbor, the small quiet town and its friendly people encouraged the impression that we had stumbled upon a perfect base for our research. There was also an excellent boatyard, Constellation Yachts, which was building large sailing boats and thus had many of the parts and much of the expertise that we needed to maintain *Tulip*.

But we had too many obligations to be able to relax and enjoy the charms of Trincomalee. Gay and Margo traveled to another east coast fishing center, Batticaloa, to investigate the dolphin by-catch there, while I plunged back into the heat and frustration of Colombo to deal with the never-ending bureaucracy. Left in Trincomalee, Jonathan and Martha made repairs to

Tulip, replenished our supplies and checked the local dolphin by-catch.

When we left Trincomalee on February 12, we were all exhausted, and not in the frame of mind to do exacting sperm whale research. As the others tried to rest below, Martha and I kept watch.

"Blow!" cried Martha an hour after leaving the harbor. "About a mile off the port bow. This is the place for whales! Shall I get some of the others up?"

"Let's wait till we know what it is. Could be a Bryde's," I replied, "The blows seem big."

There was a pause of several minutes while we sailed in the direction that we had seen the whale. Then a large, powerful blow appeared about two hundred fifty yards ahead. It was followed by a long, gray back.

Martha turned to me with an enormous smile of joy covering her face. "Hal-baby, it *is*!!!"

"It looks that way." I grinned back.

Another blow, and the back began to roll from the water. It rolled and rolled and rolled — the whale seemed to be without end. Finally a tiny, almost ridiculous-looking, dorsal fin appeared, and soon afterward huge broad flukes were thrust into the air.

"Blues! Blues!" screamed Martha ecstatically. "Y'all get on deck!"

The others tumbled sleepily out of the hatch to see the largest animal that has ever existed — the blue whale (*Balaenoptera musculus*). We had all watched many sperm and Bryde's whales during the previous two weeks, but the enormous streamlined bulk of a blue whale is unequaled in the animal kingdom. Reaching lengths of one hundred feet and weights of perhaps two hundred tons, the blue whale is very much larger than the largest dinosaur; it dwarfs elephants.

The blue whale is a mottled bluish gray. From the air it usually appears slim and streamlined. The first aerial photographs of blue whales made the swollen shapes in old drawings and photographs of bloated carcasses seem ridiculous. But recent aerial photographs have shown the blues feeding. They have enormously distended throats; the whales, using their concertinalike ventral grooves, engulf at least their own weight in seawater and strain it for edibles through their baleen. When feeding like this they look much like the old bloated image of the blue.

It was a major surprise to find blue whales off Trincomalee, for they are principally animals of cold arctic or antarctic waters, and few have escaped the greed of the whalers. Even in their prime habitat they are scarce: I have only seen three during eight seasons of research off Newfoundland. In the Gulf of Saint Lawrence, where Martha has studied them during the past two summers, perhaps the best place in the world to see blue whales, they are neither common nor numerous. What were blue whales doing off the coast of Sri Lanka, only a few days' sail from the equator?

We have sighted the blues regularly in a small area a few miles off Trincomalee, at the mouth of a huge underwater canyon, which stretches right into Trincomalee Bay. This canyon means that depths of almost three thousand feet can be found inside the bay, less than a mile and a half from land. The seas are stirred up as they pass over the rugged topography of the Trincomalee Canyon, so that the nutrients necessary for life are brought to the surface. The rivers emptying into Trincomalee Bay also bring nutrients, allowing the production of small plants, the phytoplankton. The phytoplankton are at the base of the food chain. They are the food of small zooplankton, which are eaten by larger zooplankton, which in turn are eaten by blue whales. The Bryde's whales, which seem to eat mainly small schooling fish, are one step farther up the food chain, but there may be five or six links in the chain joining the sperm whales, whose food is much larger, to the phytoplankton.

We are concerned that the flow of the Mahaweli River, Sri Lanka's largest, which provides the majority of the freshwater flow into Trincomalee Bay, is being radically altered by a huge hydroelectric and irrigation scheme, financed by foreign aid. Cynics say the Mahaweli Project is a political maneuver, in which the much-needed water is diverted from the eastern parts of Sri Lanka, where Tamils predominate, to the regions inhabited by Sinhalese, who form the majority and wield most of the political power. The official purpose of the project is to allow Sri Lanka to develop, and thereby more easily feed its growing population. But how will restrictions in the flow of the Mahaweli affect the feeding of the blue whales and the other creatures of Trincomalee Bay?

Blue whales are particular about feeding grounds, since they need to consume about four tons of plankton, which can only be of a few species, a day. Several times we have followed these whales underwater by watching the thick dark lines that appear on our recording depth sounder as the whales descend after raising their flukes. They go down to about three hundred to six hundred feet; there they encounter large swarms of plankton, which show as thick smudges on our recording depth sounder. We cannot see what takes place in this restaurant of the deep, but when the blue whales return to the surface after five or ten minutes, they often defecate. Unlike the defecations of sperm whales, which briefly stain a small patch of sea greenish brown, blue whale feces are hard to miss: many square feet of the ocean turn bright orange. One swing of a dip net can bring in about a pound of orange mush: pieces of the millions of zooplankton that the whale has digested.

There seem to be roughly ten to twenty blue whales in the area. We do not know how often, or for how long, they are present. But if their residence is at all regular, this constitutes a wonderful opportunity to study Melville's "retiring gentleman."[2]

However, like the Yankee whalers who found the blues too fast and powerful, we concentrate on the sperms.

As we motor south, we are coming abreast of Trincomalee Canyon and approaching the area where we usually find the sperms, a few miles offshore from the feeding grounds of the blues. Jonathan stops *Tulip*, and once again, through headphones, monitors the hydrophone. This time he hears the characteristic clicks of sperm whales. He changes the input to the acoustic system from the nondirectional hydrophone, with which we make our initial contact with the sperms, to the directional hydrophone, which is used for tracking them. He then makes his way aft to the pushpit, a stainless steel rail around *Tulip*'s stern, puts on another pair of headphones that hang there and turns a small wheel. The wheel is attached to a rod, which runs through a rectangular aluminum tube to a directional hydrophone five feet beneath the surface. The actual hydrophone is mounted along the axis of a cone, similar to last year's primitive directional hydrophone, but in this case the whole rotating mechanism is securely mounted on *Tulip*'s stern, and is surrounded by an acoustically transparent bomb-shaped housing. This means that the hydrophone is not subject to the noise and strain of each passing wave, and we can track the sperm whales much more comfortably and efficiently than was possible last year.

After a minute Jonathan removes the headphones and writes the direction and approximate range of the sperm whale clicks on a data sheet: 140° grade two. The loudness of the clicks is graded from one, barely audible, to five, when the whales sound as though they are clicking a few feet from the hydrophone. At grade two the whales are still several miles away; there is no point in scanning the horizon just yet.

However, Margo stirs from her bunk opposite me. "Have we got sperms?" she asks.

"Yes, but they're still quite a long way away."

"I'll get the cameras ready," she volunteers. She pulls down the aluminum camera cases from the shelf above the main bunk where they are stowed and begins cleaning lenses and checking film numbers. Gay notices the action and silently makes her way aft to the galley, where she starts preparing a lunch of crackers, cheese, fish, avocados, pineapple, mangoes and papayas. There will be few opportunities to eat when we are with the whales.

After half an hour of motoring, the clicks have increased their intensity to grade four. As we sit around the cockpit, finishing our lunch, we scan the horizon for whales. Martha sights them. "Blow! Six hundred yards, about twenty-five degrees to port." Jonathan hands *Tulip*'s tiller to me and joins the women, who are hastening below for their equipment. I steer toward the blowing sperms. Martha is the first on deck. She carries a clipboard containing several data sheets, on which she will record almost concurrently the actions of the whales, the photographs taken and other data. It is a demanding job and not very rewarding, because when the whales are most interesting the writing is most intense. However, Martha, who is usually stuck with it, has proved herself talented at making sense out of the confused and often simultaneous yells of crew members telling of sightings, behavior, or photographs taken. Following Martha come Gay and Margo, carrying cameras with telephoto lenses, one loaded with black-and-white film, the other with color. They follow Martha up to the foredeck.

Jonathan appears last. He is wearing a canvas harness, with a rope around his neck and two cameras and a small satchel containing a cassette tape recorder slung from his shoulders. He starts to climb up the mast. On reaching the spreaders halfway up, he unwinds the rope and secures it to an eye in the mast about six and a half feet above the spreaders. The rope is attached to the harness, so that now Jonathan can relax, sitting in the harness, hanging by the rope, with his feet braced against the spreaders. From this position he can, with reasonable comfort, observe and photograph the whales. In contrast to the observers on deck, he will often be able to see a little way underwater. The thin line of back that is visible to sea-level observers is rounded into a real animal with flukes and flippers when seen from above. He also has a clearer image of the whales' behavior. He can see their forms swimming just beneath the water, or the flash of white as a whale rolls to reveal a belly patch.

I am left in the cockpit. Unlike the others, I am unfettered by cameras, data sheets or harnesses. I steer *Tulip* toward the blows, but cannot resist munching on the remains of lunch, which are scattered around the cockpit.

Two misty bushes appearing almost simultaneously from the water in front of us are quickly followed by a third. Soon we see the loglike backs. A square head is thrust slowly out of the water. "Spyhop! Did you get it?" Martha asks the photographers.

"Yeah, frame twenty-seven" and "Sure, number sixteen," reply Gay and Margo simultaneously.

"One of them seems to have separated," calls Jonathan from the mast a minute later. The mist of one blow is now fifty yards from the other two.

Suddenly the huge hind part of a whale lifts from the water, then thrashes downward with the flukes slamming flat on the surface. There is a large initial splash and a loud slap, followed, a second later, by a geyser of water that has rushed into the void created by the flukes.

"Lobtail!" cries Margo, using the whalers' term for this dramatic activity. The whale makes another ostentatious lobtail as it swims away from its erstwhile companions. Gay and Margo take photographs. "That goes along with your theory, Hal!" Gay calls to me. I had found during my research on humpback whales that we often saw spyhops, lobtails and breaches as groups split up, which suggested some kind of social reason for performing these activities. The lobtail might signal the whale's "state of mind," or it might be an emphasis applied to another, probably vocal, communication. Just as we jump up and down, or wave our arms in the air to accentuate a statement about which we feel strongly, so, perhaps, the whale breaches or lobtails. These lobtails by the departing whale fit my theory, but I would like a larger sample size than two before extending it to sperm whales.

We come slowly up to the remaining pair of whales. They are moving steadily but very slowly forward at the surface. Jonathan is quietly recording their blowing patterns into the microphone of his cassette recorder: "Blow — Whale A . . . blow — Whale B . . . blow — Whale A . . . ." As we slip alongside, about twenty yards from the whales, Gay and Margo photograph the dorsal fins, so that Jonathan, many months later in his laboratory at Cambridge, may be able to answer Martha's question "Wasn't that closer one, with the notch in the trailing edge, in the big group yesterday evening?"

Jonathan has taken photographs from his mast position that show the whales lying parallel to the horizon. By making measurements on the negatives of the lengths of the images of the whales and their distances from the horizon, he will later be able to estimate their actual sizes.

Gay whispers briefly to Margo and they aim their long lenses at Jonathan's left foot, braced against the spreaders above their heads. "You, too, Jonathan," calls Gay, and he is forced to aim his own shorter lens downward at his left foot. The three left feet will form distinctive markers on the negatives. They will ease the problems of coordinating all the different data sources many months from now in the laboratory, when the events of today are reduced to a blur in our memories.

By now I have maneuvered *Tulip* so that she is hanging forty yards behind the pair of whales. Soon they accelerate, arch their backs and raise their flukes to begin their descent. Cameras click, and Jonathan calls out, "Shit!" Margo grabs the dip net, while I steer *Tulip* toward the slicks left by the diving whales. A touch of reverse as Margo flourishes the net, and we have three more squid beaks to add to our fast-growing collection of sperm whale defecations.

"Let's try BBN-ing. There may be enough breeze to sail," I call, leaving the helm to Gay and going below to the chart table.

"BBN-ing" is our name for trying to track the whales underwater using the new acoustic equipment. I turn on the depth sounder, tape recorder, preamplifiers and oscilloscope. Gay stops the engine and pays out equal amounts of the cables of two nondirectional hydrophones, one trailing from each side of *Tulip*. Jonathan has climbed down from the mast and is hauling up the genoa, *Tulip*'s largest jib, to catch the zephyrs.

On the depth sounder a pair of parallel diagonal lines appear. They are close together and represent the diving whales. I watch them descend, calling the depths into the microphone on the tape recorder. "Three hundred feet . . . six hundred feet . . . nine hundred feet."

At three hundred feet we start to hear the clicks of the diving whales; dark flecks, representing the clicks, appear on the depth sounder output. One line has now vanished; the whales have split up. To follow the whales on our recording depth sounder we must be vertically above them, because the depth sounder's transducer, which emits and receives its pings, points downward. Luckily the whales usually seem to start their dive almost vertically downward. However, when the whale reaches its "cruising depth" — perhaps eleven hundred feet — and starts to move horizontally, we have to follow it. This is when the BBN is employed. It has been hard to use for its original purpose — acoustically distinguishing individual sperm whales — but it has proved to be efficient during this underwater tracking.

The technique was developed by Jonathan and Lex Hiby, experimenting in the flooded gravel pits near Cambridge, but it is also effective here in the Indian Ocean. The two hydrophones each hear a sperm whale click, but if the whale is to the right of the hydrophones, the right-hand one picks it up first. The oscilloscope displays the clicks one above each other. The upper click from the left-hand hydrophone appears to be slightly ahead. "Port a little," I call to Gay. She turns the tiller, and *Tulip* swings until the two oscilloscope traces are exactly above each other.

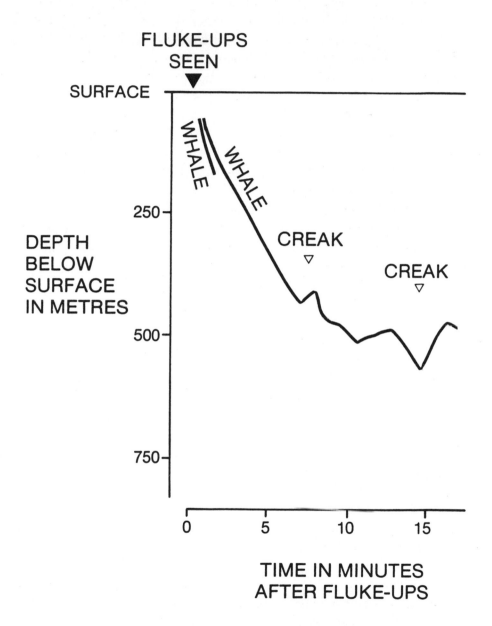

Figure 2

"Right on," I tell her, and she straightens the helm. The whale is now directly ahead and far beneath. I watch its trace on the depth sounder, as the steady clicking thuds from the tape recorder loudspeaker. The depth sounder trace of a sperm whale provides a fascinating and informative record of its dive. We can work out how fast it is descending, watch it passing through plankton concentrations, grazing the bottom, or meeting other whales underwater.

This whale's trace, which is descending smoothly (represented by a straight diagonal line), suddenly rises and then levels off over a few seconds. (fig. 2) As the whale makes this sudden maneuver far beneath us, I simultaneously hear an accelerating series of clicks that sound like a creaking door. There is a silence lasting a few seconds, and then the regular clicking resumes and the steady descent of the depth sounder trace recommences. It seems that the whale has found something.

I try to visualize the scene in the depths beneath: the whale received an echo from her regular clicking. The bearing of the echo and the time delay told her the position of the target. Other features of the returning pulse probably indicated the nature of the object. It must have sounded promising, for the whale then started swimming toward the target. As she came closer, each echo was received more quickly and each subsequent click made after a shorter delay, producing a series of clicks with an increasing repetition rate — the creaking door. Tantalizing questions remain: Did the silence indicate an attack? What was the target — a squid? Was it eaten? A detailed analysis of the creak on an oscilloscope will surely tell us more about the encounter, but much will remain a mystery until we can accompany the whale herself during her foraging in the deep.

The roll of paper that is the depth sounder output and the tape recordings of the whale's clicks are the only glimpses we have of the greater part of the life of the sperm whale. But, perhaps from these sketchy chronicles of the sperm whale's dive, we can glean something of this realm of pitch blackness and devastating pressures — the most remote of the world's ecosystems.

BBN-ing is a difficult business, and we are just learning the tricks of the trade, so that after twelve minutes we lose the whale's trace from the depth sounder. A natural pause gives us an opportunity to organize the data we have just collected. I mark the depth sounder and make entries in the catalog of tape recordings and ship's log, Jonathan hauls the hydrophones in, Gay changes the film in her camera, Margo bottles the squid beaks from the fecal sample and Martha arranges her many data sheets. But we are soon off after another cluster of blows.

This time there are five loglike backs lying at the surface. We are alongside them for several minutes while the identification and measurement photographs are taken. "The two closest to us have calluses — big mamas!" says Martha. On the crests of their dorsal fins are white, calluslike patches, each a few inches across. Research by two Japanese scientists, Toshio Kasuya and Seiji Ohsumi, has shown that these calluses are indicative of mature females[3]; they are secondary sexual characteristics, like the beards of human males. Like beards, they do not appear to serve any obviously useful function for the whales. But they are useful to us, allowing a fairly reliable guess as to the whale's sex from a simple dorsal fin photograph.

"There's also a little one," calls Jonathan from his vantage point up the mast. Sure enough, a "tiny" — twenty-foot — calf leaves its sedentary elders and swims toward *Tulip*. The calf, probably a few months old, possesses all the charm of most young animals, showing an endearing trust and curiosity. It swims steadily, with head bobbing to the strokes of its small flukes.

"Hi, baby!" cries Martha.

Appearing curious, it circles *Tulip* at a range of about six feet, while we hang over the rails, entranced. Then, having checked out this new addition to its environment, the young whale swims back to its family group. Ishmael also encountered sperm whale calves who "evinced a wondrous fearlessness and confidence . . . which it was impossible not to marvel at."[4]

Delightful as such incidents are, they are a diversion from one of the principles of our research — passivity. We are trying to understand the natural behavior of whales, not to seek out interactions with them. If we are ever to be able to communicate with whales, we must first have a good understanding of their natural behavior, especially of their relations with one another. To do this, our observation must be as passive as possible. This principle of passivity contradicts the experimental basis of most science, but I wish it could be more of a fundamental principle during our interactions with nature. If only man could try to minimize his trampling on the intricately evolved web that all life depends upon.

Having obtained our identification and measuring photographs, we hang back, staying two hundred yards from the whales. We listen through the hydrophone. There are none of the regular clicks so characteristic of sperm whales. Mostly there is silence, but then: "Click, click, click — pause — click, click."

Another whale seems to answer: "Click — pause — click, click, click, click."

And then: "Click, click, click — pause — click, click," the first whale seems to reply. These are the remarkable sperm whale "codas." They were discovered by Dr. William Watkins of the Woods Hole Oceanographic Institution in Massachusetts.[5] He named them after the term for a section of a musical composition formally distinct from the main structure. By listening to sperm whales with extensive arrays of hydrophones, Dr. Watkins has found that each sperm whale seems to have an individual coda, and that they may be used to convey information. We often hear the codas in social circumstances, when the whales are maneuvering at the surface in relatively large clusters, as is the case today.

It is calm. There is little noise from waves or water sluicing past the hydrophone to detract from the quality of our recording of these codas. But as I listen, a low whirring comes through the headphones. Its volume increases. "Fishing boat coming," says Jonathan. I end the contaminated recording and go on deck. A brightly colored thirty-foot-long Sri Lankan fishing boat is coming alongside. One crew member sits on top of the dog-house, steering lazily with his foot on a long tiller. His three laughing colleagues dressed in straw hats and sarongs proudly hold up large tuna and gruesome-looking sharks they have caught. They wave their hands in front of their mouths, as though drawing on cigarettes. For some reason the Trincomalee fishermen believe that yachts are a good source of free ciga-rettes. We try to dicourage this concept, which has cost us several good recordings, and shake our heads. Unfortunately a headshake in Sri Lanka can mean yes, or at least, as Jonathan has observed, "I don't understand what you are saying, but it sounds all right to me." The fishermen come closer; they are amazed when we appear angry, yelling, "Go away!"

They leave cigaretteless and perplexed. We try to signal that they should give the whales a wide berth. But despite our entreaties, they motor at full speed just past the animals we were watching. The whales startle and dive. Until we learn to convey simple information to our fellow humans, what chance have we of communicating meaningfully with whales?

8

EAST OF SRI LANKA

February 18 - March 11

A WARM BREEZE FROM FARTHER EAST spreads a sparkling moon over the skin of the ocean. Perched on the edge of *Tulip*, I let the dark waters flow through my troubled soul. Longing for home in Newfoundland and the woman I left there nearly three months ago, I search the seas for a link to them.

The wind turns the water here above the Trincomalee Canyon, on the edge of the deep. Like a plow it brings up cool water, which many years ago had left the antarctic ice; the deep water, rich with nutrients, tumbles upward toward the surface, enriching these tropical seas. Far to the south, that same ice begins its annual winter growth. It sends its chill tongues northward, throbbing beneath the seas to the Trincomalee Canyon, to other cavernous edges of other oceans.

The deep hiss of a whale blow whispers through the warm night.

Oh, that I could dive over the rail to become one of these whales; to swim down — darker, cooler, stiller — watching phosphorescent flashes of other life, hearing the muted roar of the surface above. Downward, I feel the layers, each of different origin, with its own taste and feel and life. Down to the southward countercurrent that we have used for centuries, there to swim rhythmically — click for food when hunger comes, up to breathe each hour or so. South to the ice, where I would turn to the northwest, dive deeper — a new current to take me home to Newfoundland — dark. Across the equator I hear the clicks of old companions, the reverberations of familiar haunts. On, northward — seamounts hiss with life at either hand. Slowly the roar of the shelf breaks through. The Grand Bank is storm driven in March — any antarctic water climbing its sides is quickly lost in chaos, beneath scudding sheets of cloud and a solitary, skimming, searching bird. The northern ice drives down from the Labrador. Cradled on the Grand Bank is the battled Rock of Newfoundland. Clouds wash, ice beats, the Rock endures. Wooden houses cling to the cliffs, paint peels — home, where my love waits.

Martha sits down beside me. "How you doin', Hal-baby?"

I can only mutter. "What an awful day. I hope the engine keeps going . . . the moon . . ." Why can't I pour out my exhaustion, homesickness, sadness and loneliness? There is no more sympathetic ear than Martha's. But I am bottled up where only the ocean can soothe, and I remain silent. So, with a "Hang in there. We're all with you," Martha moves forward to Gay, who has also been looking to the waters for strength. Soon their murmurs float aft on the breeze. From astern comes the full-throated laugh of Roger Payne, who has finally made it on board *Tulip*. He and Jonathan discuss ideas and wild schemes for the future.

Alone, too tired to contemplate any future but the peaceful release of Newfoundland, my only immediate desire is that the oil-pressure gauge on the engine remain steady. The days have been too full of the whales, our lives and work, for another night of contortion around the engine, smeared with its hot, black oil. Three weeks ago we were near our limit as we struggled, in happy companionship, to follow and study the whales. But the pressures have intensified and cracked the fine edifice we had, built.

It was at the Symposium on the Marine Mammals of the Indian Ocean, held in Colombo at the end of February, that we first felt the effects of the strain. Arriving with film and videotapes of the blue whales we had just discovered off Trincomalee, we were a center of attention, and found ourselves beset by demands from many sources. In addition to the presentations that we and others gave, the scientific discussion and exchange of information, we found ourselves drawing up plans for Sri Lankan whale research at the request of N.A.R.A., the newly set-up Sri Lankan marine resources agency. We advised on research techniques and whale tourism. We even became embroiled in a debate on the best techniques for conserving dugongs; N.A.R.A. was keen to establish a captive population, but we and others believed education and enforcement of regulations more important. Friends from earlier stages of the study (Sidney Holt, Roger Payne, Pieter Lagendijk and Lex Hiby) were attending the conference — there were so many ideas and plans to discuss, between the lectures and receptions and visits. And we were constantly being approached by people who wanted to join us on *Tulip*.

There were also demands peripheral to the symposium. Margo arranged radio and newspaper interviews; Gay spent many hours trying to goad bureaucrats into allowing her dolphin skulls to be sent to the United States; Martha bought food; Jonathan and I, engine parts. Unexpected obstacles appeared: Banno Lanka's boatyard had been seized by creditors (something to do with the barge that sank during launching), and with it Gay's buried dolphin skulls — more rounds of bureaucracy before they could be disinterred.

The most disturbing aspect of the time of the symposium concerned the film that W.W.F. had arranged to have made about the whales and our work. We expected to find the crew in Sri Lanka, but there was no sign of them. We made many international phone calls to try to find out what had happened, but all that became clear was that the plans of this filmmaker should not be trusted.

The symposium ended with a day trip from Trincomalee to see the whales. On board *Tulip* we took Lex Hiby and Pieter Lagendijk, our two Dutch shipmates from last season. Pieter was thrilled to finally see and film the sperm whales he had missed by one day off Oman the previous year. Following us came a Sri Lankan Navy tug loaded with N.A.R.A. officials, their guests, participants in the symposium and a Sri Lankan television crew. The blue whales they saw were right outside the harbor, and cooperative. All those on board were stirred by their first sight of the vast animals, and the Trincomalee whale-watching industry was born.

Finally, on March 2, we left behind the circus that land had become and headed for sea. With us came Roger Payne, who was replacing Margo for this two-week trip. Although I have known Roger Payne for eight years and had worked in his laboratory in Lincoln, Massachusetts, this is my first opportunity to be at sea in his company. After I had analyzed his data and talked to those who had been with him in Argentina, it was clear to me that Roger has an extraordinarily intuitive feel for the lives of whales and that he is dynamically creative in his approach to understanding them. Ideas and schemes stream from his agile brain, often with unexpected consequences. One inspiration has led to our cockpit becoming full of rocks. We had been discussing sperm whale creaks as a form of echo-location. "If only we could get the whales to creak off some known object," said Roger. "Maybe if something was thrown overboard when we were above a whale, and we could track the object and the whale on the depth sounder, we could see if a creak occurred when they came together." So one night, when we had taken shelter behind a small island, a shore party made off with twenty large rocks. They now await the perfect opportunity to be dropped.

Roger's conceptual framework for viewing whales is rather different from mine. He treats the large and powerful animals with considerable respect and caution, whereas to me the potential for danger from a whale is so minuscule compared with terrestrial hazards — trying to cross the street in New York City, for instance — that I have never feared them. This difference of attitude surfaced three days ago when Martha yelled, "Breach! Eight hundred yards off the starboard beam." Our eyes followed her pointing arm, where the bulk of a sperm whale soared fleetingly through the air. "There she goes again," cried Martha. "Let's get over there!" We started the engine, and I began to steer *Tulip* slowly toward the breach. I do not particularly believe in rushing toward every breach in order to try to take the perfect photograph of a leaping whale. However, there were no other whales near us, and it would be interesting to draw a little closer to the breacher in order to see if it had companions.

But Roger, to my surprise, advised caution in our approach. He believes that breaching may often be an aggressive display and could be directed toward a research vessel. Jonathan shares Roger's belief that whales may be dangerous and should be treated with care. In contrast, I have a firm conviction that, although they can act aggressively toward one another, whales are basically harmless to humans if the humans do not act aggressively toward the whales. I think that in general they breach in order to communicate information to other whales, and if they injure people or boats, it is only by accident or after exceptional provocation. Martha, another reckless soul, backed me up. The vigor of our arguments almost matched the strength of the whale's breach.

This is an old debate. In 1839, Thomas Beale refuted those, like the great French zoologist Baron Cuvier, who declared: "There is no animal in creation more monstrously ferocious than the sperm whale." Beale himself regarded the sperm whale as "a most timid and inoffensive animal as I have before stated, readily endeavouring to escape from the slightest thing which bears an unusual appearance."[1]

Current images of the sperm whale as ferocious are largely due to *Moby-Dick*. Melville declared that "the sperm whale is in some cases sufficiently powerful, knowing, and judiciously malicious, as with direct aforethought to stave in, utterly destroy, and sink a large ship."[2] However, the wheel is turning, and in current popular literature whales have become "gentle giants." In *Tulip*'s cockpit three days ago this same debate grew heated: Roger and Jonathan advised caution when sailing with the sperms, while Martha and I defended their inoffensive natures. Gay, open-minded about whale behavior, stayed neutral, but finally joked, "Perhaps it was just playing."

"As for play . . ." and "I'm not sure that one can say that . . ." Roger and I began, and another discussion — on whether whales play — was born. However, this time we found that we more or less agreed.

A person seeing an animal perform an activity for no obvious reason will often say, "It is playing." Thus "play" becomes a dump for the disposal of otherwise inexplicable behavior, and in general, the less animals are understood, the more they are said to play. Breaching, a not immediately obvious activity for an aquatic animal, has often been so designated. In both Roger's studies of southern right whales and my work on humpbacks, detailed observation has suggested good reasons why animals might be performing activities that had previously been called play. For instance, Roger has suggested that young right whales may breach on their mother's backs to persuade them to roll over into a more favorable position for suckling, or that female right whales may raise the whole rear part of their bodies out of the water to avoid being mated by undesirable males. On Silver Bank in the West Indies, I had seen humpbacks charging about apparently randomly. At first they seemed to be playing Follow the Leader. However, close observation finally showed that they were males boisterously competing for access to a central female.[3]

Scientists do concede, however, that there is a class of

behavior called play, and although there are difficulties in defining it, serious research has been carried out into its nature and development. During our work on whales, we have borne in mind the writings of those who study play, and the whales do indeed sometimes seem to be playing. It is easier to differentiate play from the more immediately serious aspects of life when the play involves an object, so most of our observations of play concern objects, although the less distinguishable social play may be equally as common. In Argentina, Roger Payne has seen right whales apparently playing by passing pieces of seaweed over their backs, and off Newfoundland we saw a solitary humpback rolling a plastic bucket. The whale, a male, finally turned upside down, so that its belly was grazing the surface, and then tried to urinate into the bucket that was floating alongside. There is no conceivable function for toilet training in whales, so we can reasonably assume that the whale was playing. Sperm whales also seem to play: on February 4 we saw one pushing a stick along the surface with its head. Breaches may occasionally be play, but to be certain of this, we would first have to rule out all other possible reasons.

Besides stimulating discussion and casting out new ideas, Roger has thrown himself enthusiastically into the life on board *Tulip*; he stands watch, takes notes, cleans and cooks. But I miss Margo's steady reliability — the least noticeable crew member, she was also perhaps the most important.

The research, which was always hard, has become even more complex with the addition of Roger's schemes and new approaches. The drain of our last spell on land for the symposium has been augmented by the increasing unreliability of *Tulip*'s engine. My friendships with Gay and Jonathan have deteriorated as the two of them have grown closer to each other, and I seem unable to accept Martha's freely offered sympathy. I feel wrung out.

The research ideals that I believe in — particularly the rigorous collection of standard data — seem to have been eroded by the general exhaustion and the flood of new ideas. I doubt myself and my abilities, as well as my role in this study. My mind seethes. I pushed the *Tulip* Project here, gave my experience and energy, but now the study is established — and it belongs to Jonathan. So, I suppose, a mother feels when her child is ready to leave home: she has done her part in raising him and has to learn to let go, even though his future life is not aimed quite as she had hoped. Her achievement is that he lives vigorously and well, not that he lives in her mold. There is some satisfaction in this thought, although it is difficult to withdraw gracefully.

There is one development I have had only a small part in, but that has provided a great deal of excitement and drama: we have been attempting to swim with the whales. The scientific rationale for these efforts is to try to obtain better views of the sperms — to watch their behavior, to check on their sex by examining the genital slits. From the deck of *Tulip* we can see little more than the loglike backs — hardly a glimpse into the life or nature of the whale. It would be analagous to seeing only the narrow strip of skin around a human's spine and trying to deduce the behavior and personality of the subject. From Jonathan's position up the mast he can see a little way underwater — if the whales are close, if the sun is high and if the sea is calm. But it is little more than a glance. We would like to watch the whales underwater, see them in their element, with no barriers.

It was Jonathan, an experienced diver, who planned this venture. As with everything he does, he has proceeded carefully and rationally. First he investigated a possible technique by which we might view the whales underwater, without diving. He attached an underwater video camera to *Tulip*'s keel, hoping that we could watch the whales on a monitor in the cabin. If successful, such a system would be logistically simpler than having crew members actually in the water. We would not have to worry about their safety or stop to let them reboard the boat.

Although we caught a few glimpses of fleeting shapes, it was soon apparent that this system was insufficiently flexible to gain any real impression of the whales. We would actually have to enter the water ourselves. From then on, Jonathan and the other crew members began to slip into the water when sperm whales were close. Their first efforts were tentative and usually unsuccessful; no one had any idea what to expect when swimming with sperm whales. Finally, Gay saw a whale, about thirty-five feet long, heading toward her.

"I was stunned and confused," she told us later. "With one hand I was trying to hold on to *Tulip*, while with the other I fumbled with the camera. I tried to collect my wits in order to photograph the shape looming toward me. But I kept wondering whether I looked anything like a baby giant squid. I gave up on the camera and just waited. The whale glided past me. It was so slow and effortless! It turned its head directly at me and the boat, and I had this intense feeling *I* was the one being studied. Slowly the whale turned its belly toward me and floated past."

"Were you able to photograph its genital slits?" asked Jonathan.

"Ah, if only I had been cool and calm it would have been easy," Gay replied.

"But what did it feel like?" Martha asked, fascinated.

"I was left thinking, so this is what a sperm whale really looks like! After all these months of only seeing the whales from the deck of *Tulip*, I realized that I had never truly seen them before. This was the real sperm whale."

The others have experienced the same awe when swimming with the whales, and sometimes it has been hard to stop the whole crew from jumping in the water — such has been their enthusiasm for these encounters.

We have gradually worked out a routine. A long rope is trailed behind *Tulip*, and the swimmer, wearing mask, snorkel and sometimes flippers, climbs down *Tulip*'s stern — difficult when wearing flippers. He or she hangs on to Clarence, the self-steering gear, and drags behind in the water. On nearing the whales, *Tulip* is slowed as much as possible so that the swimmer can approach them, keeping hold of the long rope. In case of an emergency the swimmer can quickly pull himself, or be pulled, back to the boat. And if *Tulip* needs to move on to keep up with the whales, we do not have to wait until the swimmer has returned. However, the swimmers have a natural tendency when at the limit of the rope, and in the presence of whales, to let it slip through their fingers.

Using these techniques, Jonathan has had some fine views of the whales underwater; by making short dives, he has sometimes managed to swim underneath whales so that he can view their genital areas and thus definitively determine their sex. So far, all those that he has found to be females have had the pale dorsal fin callus, whereas the males have not.

A few days ago, Martha teased the "ferocious nature" of the sperm whale when, while swimming behind *Tulip*, she was approached by an adult. She fearlessly dropped the rope and swam steadily, directly toward the whale. As the two approached each other, those of us watching from *Tulip's* deck wondered who would give way first — a game of "chicken" on the high seas. Martha kept right on swimming; when the two were a few yards apart, the whale veered off, having apparently come close enough to this alien form.

But Gay has had the most delightful underwater encounter with the sperms. On March 7 she entered the water beside two animals that were rolling around each other. By this time she felt a veteran and much more at ease.

"The two whales began to descend facing each other," she reported. "Their heads came together and their mouths touched as their bodies streamed upward."

"They were kissing!" interjected Martha excitedly.

"Maybe they were. Watching them gently fall below me in such an endearing manner made me laugh and gurgle myself silly. They *are* loving and gentle. There's no vengeful Moby-Dick!"

These experiences gave us confidence in the safety and value of swimming with whales, but Jonathan continued to advise caution. We only entered the water in calm weather when the whales were close, and Jonathan was strongly against letting go of the safety line, despite the temptation to swim up to the whales. We will be more careful after today.

A few hours ago, in the early afternoon, Gay and Martha were trailing behind on the rope, when they glimpsed some whales off to port. They dropped the line and swam closer, but the whales moved out of visual range. The swimmers were about a hundred yards from *Tulip*, who was lying "dead in the water," when Gay raised her head and yelled, "Shark!"

Gay was being approached by a shark of about her own size. She behaved bravely and correctly. As the fish moved in, she stayed calm and turned toward it. It continued its steady approach to within a yard. Gay then kicked it in the snout (a particularly sensitive area) with her flipper. Then, instead of rushing for the boat, she stood her ground and kept her eye on the huge fish as it skulked away.

It seemed to take hours to get *Tulip* started, turned around and Gay and Martha back on board. Everyone was distraught — Jonathan was furious that I had allowed them to leave the rope. I felt the weight of my responsibility as skipper. We all were imagining "What if . . . ?" Sharks are not supposed to be dangerous around Sri Lanka, but few had experience of them in deep waters, away from the coral reefs where they can find an easy meal.

Tension and concern hang in the air tonight.

Roger calls quietly to me, "Hal, I'm afraid that the oil pressure is dropping."

"Okay, turn the engine off. I'll try changing the oil pump." I take a deep breath of the sparkling ocean air and go down below.

9

NORTHEAST OF SRI LANKA

March 12 - April 8

WHILE THE REST OF US WERE AT SEA with Roger Payne, Margo gave talks about whales, *Tulip*'s work and the Sri Lankan marine environment to several schools and one youth group in the Colombo area. Her audiences were enthusiastic: they were thrilled to glimpse a previously unknown part of their world. In the long run education is probably the key to preserving the marine environment. If just a few of her students retain a fascination for the ocean that will temper their future actions, then Margo's teaching may have been the most important part of the whole *Tulip* Project.

Conversely, the success of her teaching gave Margo a greater sense of her part in the project. She returned to *Tulip* refreshed and eager just as we were at our lowest. The powerful experiences, long hours and provocative ideas had stressed us all, but especially me. Despite the intensity of the research, the sea usually forms a soothing relief from the frustrating battles of land. The rhythms of our seagoing lives, our companionships and the encounters with the whales give structure and joy. But this time the rhythms were masked by engine problems, we disagreed with and upset each other, and had our most intense experience with a shark, not a whale. Even Martha's exuberance could not resolve the tensions that had started to develop. But now Margo substituted for Martha, whose turn it was for a rest ashore, while Roger was replaced by a Sri Lankan, Cedric Martenstyn. In the place of *Tulip*'s two most spirited and outgoing crew members, we were joined by two quiet and capable companions. The exacting intensity of life diminished, and the crew slowly regained composure.

Cedric is a Sri Lankan of Dutch origin, a member of the small European "burgher" population that lingers from colonial days. In his late thirties, he has worked at several jobs, the most rewarding of which was diving off the Sri Lankan coast. He has dived for wrecks, for coral reef fish and to clean ships. During his many adventures he has gained a deep knowledge of the coast, and he shared his yarns with us on quiet evenings. Although he had little experience of whales, he taught us much about the waters we were sailing.

"There's a calf right here!" cries Margo. An indistinct dark shape under the capricious waves consolidates as the whale rises, then suddenly transforms into a solid, breathing animal. The little sperm whale rocks its head and heaves its back above the water as it plows steadily southward — the direction in which *Tulip* pursues the clicks of the sperms who are feeding beneath. Some light parallel scars etch the calf's smooth case behind its blowhole.

"It's 'Scratch-face' again," says Gay, "and there's a bigger one just beyond."

The transformation of dimpled blue ocean, to dark ambiguity, to living, blowing whale is now repeated on a larger scale as a full-size adult surfaces ten yards beyond the calf.

"A triangular fin with a notch on top," notes Jonathan.

Scratch-face has changed companions. Just ten minutes ago he was with a whale who had a rounded fin and a callus.

This is the second time today we have seen a young sperm

whale change companions within a few minutes. Initially we assumed that, as with humpback whales, the adult who closely accompanied a particular calf was probably its mother. But soon we noticed that some of the companions did not possess the telltale dorsal-fin callus of the adult female sperms. Jonathan was able to confirm by inspection of the genital area when diving that, on at least one occasion, a calf's companion was a juvenile male.

"There's another calf, about one hundred yards farther over," calls Margo again. "It has lots of remoras around the dorsal fin." Most of the small sperm whale calves carry hitchhiking remora fish. The foot-long, gray, shark-shaped remoras have large suction pads on the tops of their heads, which they use to attach themselves to larger fish, boats or whales, picking up a free ride. They can leave a particular host at will, and we have watched them transferring between *Tulip* and the whales. Adult sperms do not carry remoras, but blue whales of all sizes do. This suggests to us that remoras cannot endure the conditions at the depths to which adult sperms regularly dive, and that sperm whale calves, like blues, are restricted to the upper layers of the oceans. Gay thinks that although remoras may perhaps be feeding on external parasites of their host and thus benefiting the whale, it is the remora that receives the major advantages from the relationship. The whale finds feeding sites and takes its remoras to them, and while providing free transportation, keeps water flowing over the remoras' gills, aiding their respiration. Cedric immediately knew the remoras that rode on the blue whales and baby sperms; he had seen the same species many times hitching rides on sharks and groupers.

We often see small calves trundling along at the surface in the direction of the clicks of adult sperm whales beneath, much as we do on board *Tulip*. Sometimes they lie almost still or turn circles as they wait for the return of their elders, but on other occasions, like today, it is a struggle for both the calves and *Tulip* to keep up with the clicks. This could be due to the speed of the adults traveling down below, or more probably, to the effects of layers of currents. If the surface water is moving north at two knots and the water fifteen hundred feet down is moving south at two knots, then a calf at the surface would need to swim southward at four knots just to keep up with stationary adults beneath.

The calves are members of groups of twenty or so adult females and juveniles, and when most of the larger whales are feeding at depth, one of them is usually left behind at the surface near or beside the calves. Is it "baby-sitting" — keeping an eye, or more probably an ear, on the calves in case they lose their way or are attacked by orcas, also called killer whales (*Orcinus orca*), or large sharks? If this is so, then today's observations of calves switching companions means that the members of a large group of females (as well as the juvenile males in the group) baby-sit for calves who are not their own. This may be important for a mother who would otherwise frequently have to leave her offspring unattended at the surface while feeding at depth.

Scientists call this type of behavior "alloparental care"; that is, care of young by an animal who is not the parent. It is characteristic of advanced animals living in highly social groups where long-term reciprocal relationships have had the opportunity to develop. Evolutionary theory predicts that animals are only likely to care for others' offspring if they are closely related to the calf, or know there is a good probability that the favor (or some comparable favor) may be returned in due course.[1] We speculate that having reliable baby-sitters available may be a major reason why female sperm whales may form stable groupings. These long-term relationships between females do not seem to be present in any of the other large whale species, which do not dive to great depths for long periods to obtain their food and thus can stay close to their calves.

That calves might sometimes become confused if left alone at the surface was suggested by an observation we made a few weeks ago just off Trincomalee. We saw a sperm whale calf trudging along valiantly at the surface in pursuit of a fast-moving adult. But on approaching, we noticed that the larger whale, two hundred yards ahead, had a straight powerful blow. It was a Bryde's whale! Of course, given the fearlessness of sperm whale calves, it might have been simply inquisitive about the sleek Bryde's, rather than confused about its identity.

Apart from baby-sitting, alloparental care in sperm whales might take other forms. Females could assist in the instruction of one another's calves in feeding or other matters, although we have no evidence for this, and it would be difficult to obtain any. Another possible form of alloparental care was suggested by an observation I made while swimming with the whales.

On April 5, after a morning of frustrating battle with diesel mechanics while we hove-to at sea, I finally changed the oil filter, and the engine ran. In my glee I carelessly hurled the old filter overboard, splattering Gay with pitch-black engine oil. She was not pleased, but nothing could dent my happiness. We motored for two hours and then found a large group of sperm whales at the surface. They were twisting and maneuvering into unusual patterns, including a star. In this formation, their heads were together at the center, but their bodies radiated out in different directions. The Japanese scientist Masaharu Nishiwaki has also observed sperm whales taking up this formation; he described it as "like a marguerite flower," or daisy.[2] But what is its function? Sperm whales are large animals, and their frontal region, which contains the mouth, eyes, ears, echo-location apparatus and flippers, is much more sensitive and responsive than the rear part of their bodies. Therefore, if several sperm whales wish to interact closely at the same time — all being close to the sensitive parts of each other — then the "marguerite" formation is a sensible formation, much like the American football players' "huddle."

Everyone wanted to go swimming with these whales: during the afternoon the crew took turns in the water. Jonathan managed to identify the sexes of a few animals, and later Gay and Margo had an unusual surprise when they tried to count the sperm whales swimming around them. Usually six is about the limit that any diver can see, but on this occasion, their count had passed twelve, and still there were more whales gliding and twisting at the limits of their vision. A few came closer, and the enigma was solved: most of the whales in sight were not sperms. Although similar in general shape, these new arrivals were only about one-third the size of adult sperms, and they had large dorsal fins and white anchor-shaped belly patches. The sperms had been joined by a school of pilot whales (*Globicephala macrorhynchus*). The more agile pilot whales swooped around the sperms, appearing to tease them. Like sperms, pilot whales principally eat squid, but only relatively small ones. Unlike orcas, pilot whales are no threat to larger cetaceans. The sperms seemed to react like adult humans who, preoccupied with "the serious things of life," are being harassed by energetic, inquisitive children. They appeared to be trying to brush off the pilot whales with a flick of their flukes, or swim out of the rambunctious pilots' path. The pilot whales soon grew tired of these uncooperative playmates and moved on.

Finally, as the tropical sun dived for the horizon, the rest of the crew had their fill of swimming with the sperms; their limbs were tired and their senses overwhelmed by what they had experienced.

Martha turned to me. "Hal, why don't you go in?"

"Oh, I don't know. I'm a sailor, not a diver." Until now I had not tried to swim with the whales.

"Come on, Hal-baby. It's easy," she persisted.

So I put on a mask and snorkel. In order to keep the procedure as painless as possible I left flippers behind. I grabbed our underwater camera and descended from the stern. Once in the water, I clung to Clarence with one arm. One hand fiddled with the mask to clear it of water, while with the other I grabbed my shorts to prevent them from being dragged off by the motion of the water. I caught glimpses of the strange shapes that sperm whales form underwater; a square blocklike forehead transforming into a slim, streamlined wedge by a forty-five-degree turn. Although *Tulip* was moving slowly, the drag was powerful. I felt as though I were being towed through treacle behind an express train. Frustrated, I let go of Clarence, forgetting the rope that I was supposed to grasp. *Tulip* was left behind. I no longer knew she existed. I was part of the sea and the whales that were hanging like living monuments in this realm of blue. I swam toward them slowly, calmly. There was nowhere else. I am usually an awkward swimmer, struggling for the next breath, but now the world of the whales inhabited me. It was easy; it was my element. The water was so clear that life, too, became clear. There were only the whales and me — nothing else. Two whales glided toward me. As they approached one extended a small flipper, gently touching his companion — for reassurance? Coming closer, they separated slightly and then the flippers were extended again, this time to steer. The whales' motion was effortlessly transformed from the horizontal to the vertical, and without a beat from the flukes they streamed away beneath me.

A female hung vertically in the water, her belly toward me, her head breaking the surface above — a spyhop for those on *Tulip*. Beneath, two small calves lying side by side nuzzled her genital area. Were they both nursing?

How long my dream/reality lasted I do not know, but I was awakened by calls to return to the boat. I ignored them; the calls were from a different world. The cries became more insistent. I was distracted, and lost sight of the whales. I swam back to *Tulip*, feeling as though I had just returned from a long voyage to another planet. The whales had taken a new form in my consciousness. No longer just a mysterious organism to be studied as objectively as possible, the sperm whale is a living, touching being.

My first swim with the whales also contained one potentially important observation. Were both the calves that I saw suckling from the same female? If so, at least one of them was not its offspring, as they were of a similar size, and female sperm whales probably never bear more than one young at a time, approximately once every five years. Like "baby-sitting," nursing the calves of other females would be a form of alloparental care. It is found in a few advanced mammal societies, such as elephants and some monkey species. If there were several females in a calf's maternal group from which it could suckle, this would assist each female, as the load of nursing could be shared. Whenever a mother was temporarily sick, or had poor feeding success, then her companions could, for a while, bear the major burden of feeding her calf.

Since the break for the symposium in late February, most of the calves we have sighted have been small — thirteen to sixteen feet long. In contrast, we saw no such small calves during the first five weeks of the study. Are the calves being born here? This would be the expected time of year for animals based in the Southern Hemisphere, according to the data collected by the whalers. Northern Hemisphere animals, six months out of phase, would give birth around October.

A particular pleasure in the sighting of "Scratch-face" and the other tiny calf that carried so many remoras is that these are the two calves we swam with three days ago about fifteen miles to the southeast. Since then, *Tulip* has been continuously following their group. During the night we listen for clicks and track them with the directional hydrophone; during the day we scan the horizon for their blows. We grow to know the individuals in the group; we watch their associations and behavior. Particular whales start to show personality traits: "Scratch-face" is especially friendly toward *Tulip*; a distinctive adult with a fluke-tip missing (perhaps from an encounter with a ship or orca) rolls sideways as it dives. Every hour we record their sounds, correlating codas with socializing and creaks with feeding. We are beginning to integrate *Tulip* into the life of a group of sperm whales. Our greatest fear is that we might lose them.

Each morning comes a moment of truth. As the sun rises we maneuver *Tulip* until we are surrounded by clicks. Then we watch carefully until we spot a blow, perhaps three hundred

Jonathan and all his equipment, suspended halfway up Tulip's *mast.* (A. Alling)

Hal at the chart table in front of the BBN machine, used for tracking whales underwater. (J. Gordon)

Sri Lankan fishing boat approaching Tulip *and the whales. The crew wants cigarettes — preferably American cigarettes.* (B. Coleman: A. Alling)

Small whales, like this one, sometimes seemed curious, and would approach Tulip *briefly before turning away.* (V. Papastavrou)

Martha Smythe cooking (with garlic!).
(A. Alling)

Margo Rice. (A. Alling)

Remoras cling to the flukes of a diving blue whale. (A. Alling)

Eye to eye with a sperm whale. (H. Whitehead)

Caroline Smythe. (L. Weilgart)

Linda Weilgart. (H. Whitehead)

A "spy-hopping" sperm whale displays its jaw. Is it being aggressive? (T. Arnbom)

A new-born calf, with its umbilical cord showing, swims up to the pinging depth sounder mounted on Tulip's *hull, while an adult hangs nearby.* (B. Coleman: P. Gilligan)

A sperm whale raises its flukes at the start of a long dive. (Flip Nicklin © 1984 National Geographic Society)

A sperm whale swims down to the depths where it feeds. (Flip Nicklin © 1984 National Geographic Society)

yards away. Cautiously, *Tulip* approaches until we see two long backs awash. We stare at the dorsal fins through binoculars. Are they? Yup, these are the guys we were with yesterday afternoon! Relief — we did not lose "our group" in the night.

This is my kind of research — an enduring struggle to stay with whales and to keep up standard, regular data collection. As the pile of data sheets and rolls of film grows, so our knowledge of sperm whales increases. And the longer we can stay with a particular group, the deeper we can look into their lives.

The whales' daily routine of feeding, resting and socializing becomes another harmonic in our seaborne lives, adding to and interacting with the other cycles that sway us: the phases of the moon, the circling sun, the swells and waves. To these we add our fortnightly returns to port in Trincomalee, five-day cooking schedule, three-hourly watch routine and hourly tape recordings of the whales' sounds. Within their diurnal rhythms, the whales dive for an hour, blow every fifteen seconds and click twice a second. As we spend longer with the whales, the time scale on which we can appreciate them expands. While we begin to absorb their daily life, I look forward hopefully to the time when we can sense their annual migrations, or even the five-year breeding cycle of the females. The key to savoring this existence is feeling these rhythms, following them and appreciating the unimportance of sporadic, intruding distractions.

There have been many such distractions to keep us from the whales during the past three and a half weeks. The engine has become less and less reliable. Some chronic problem with the oil system festers deep in its iron guts. I only hope it can hang on another ten days until the end of this season, after which the tired piece of machinery can be removed and completely overhauled. But it is not only the engine that is in need of repair. The mountings for the all-important directional hydrophone have grown fragile from the stress of nighttime encounters with fishing nets. Sri Lankan fishermen work mainly at night, setting long drift nets that stream from their small boats. The nets, which are made of unbreakable artificial fiber and reach several feet down from floats at the surface, are unmarked by lights, although the fishing boat, to which one end of the net is tied, may sometimes have a small fire burning in an old oil drum. The nets are supposed to catch tuna, sharks and other commercial fish, but they also entrap dolphins, turtles and, occasionally, *Tulip*. As we sail along at night, our accustomed motion will be suddenly interrupted: sometimes *Tulip* slews to a halt as the net catches on her keel; at other times there will be a loud crash as it slips under the keel to snag on the directional hydrophone, which is then pulled off. Several hours of daylight are needed to remount it.

As the year turns, the sun swings overhead and the heat builds. The strong sunlight has weakened the stitching on *Tulip*'s sails, so that, although they experience little stress from the generally light winds, the seams have begun to split. This has meant long nights spent hand-sewing the sails back together.

Exhausted by the demands of our work at sea, we have found it increasingly hard to deal with the impositions placed upon us from the land. And no trial was more dreaded than the arrival of the film crew. The colorful American crew arrived in Trincomalee on March 23 to make a TV film publicizing the whales, W.W.F., our project and Sri Lanka. It included two experienced underwater cameramen with tales of sharks and James Bond movies, an ex-mercenary and the fast-talking New York–based producer, bragging and promising.

When the film crew first arrived *Tulip*'s engine was out of action. The film crew used the time to hire a local diving boat and go out of the harbor to film blue whales. The underwater cameramen used an outboard-motor-powered inflatable Zodiak boat to approach the blues and then, when close, tumbled in beside them. Martha, whose work in the Gulf of Saint Lawrence mainly involved steering Zodiaks among blue whales, offered her services as driver. In two days, with luck, together with their own abilities and perseverance, as well as Martha's skill, the cameramen had managed to obtain some remarkable footage of the vast animals swimming underwater.

Two days later we had *Tulip*'s engine working, and it was time to try to film the sperms. We intended to lead the film crew, who had chartered their own yacht, to an area about twenty miles offshore from Trincomalee where we had regularly found groups of female sperm whales with their young. However, it soon became clear that their overloaded and underpowered boat would never make it out against the strong currents. So we turned and headed north, sailing with the current, to a closer area where we had often seen smaller single sperm whales, but never family groups.

The film crew were in luck that day: within a few hours, we had found a large family group. Soon the two cameramen were in their Zodiak, trying to enter the water beside the whales. But, unlike blue whales, the sperms were upset by the fast and noisy Zodiak, and would dive as it approached. So we towed the Zodiak quietly behind *Tulip* and, after a few unsuccessful attempts, managed to drop the photographers in the water near the whales. Both strong swimmers, using their flippers they powered quickly up to about six whales, including a calf, who were lying together at the surface. As the swimmers entered the cluster, the loglike tranquillity of the whales was transformed into turmoil; flukes thrashed and boxlike heads were thrust above the surface. We could not see the cameramen.

"They're being eaten!" we feared. Images of Ahab's battles with Moby-Dick began to flash through our minds. But as we hurried closer in *Tulip*, the photographers reappeared, shaking their fists in triumph and yelling, "Far out! That was *so* incredible!" The whales had at first startled and then become curious, swimming right up to the cameramen, who had dived to obtain a better vantage. As the whales approached, sometimes coming to within three feet of the cameramen, the calf turned somersaults about the adults' heads. The cameramen held their ground and filmed until the whales slipped into the deep.

They now had their footage. After filming a sequence of *Tulip*'s spinnaker, Wuff-Wuff, against the sunset, the film crew turned their boat for Trincomalee. The producer was going to put together his film from two days' work with the blues and one day with the sperms. The filming had proved surprisingly painless for us, and even interesting in its technicalities, yet I could not but feel that the whales were being treated rather superficially. The film will show their form and grace, but not the life of the sperm whale, which is lived over scales much longer than the few seconds the cameramen were with them.

As evening approaches we have spent four days following "Scratch-face" and his companions. May it continue.

"There's a pair of sperms at five hundred yards, thirty degrees to port," calls Jonathan from the spreaders. We motor over, but before we can get there, two pairs of flukes briefly reach for the setting sun. The afternoon, which seems to be the most restful and social time of day for the whales, is past, and now the sperms are starting their nightly feeding.

Gay checks the light meter in her camera and calls, "The light's pretty much gone now."

"Okay, let's pack it in for the day," I reply, stopping the engine.

Jonathan descends the mast, and with Margo, Martha and Gay makes his way below with the cameras and data sheets. It takes twenty-five minutes for them to put all the equipment away and sort the data collected. While the others are busy doing this, I keep an ear on the clicks through the directional hydrophone. Eventually Gay comes up. "Thank you, Hal. I'm officially on watch. Where are the whales?"

"Bearing about 230, but they're pretty close. I don't think you'll need to sail much."

"We don't want to lose them, do we?" asks Gay with a smile, easing the mainsail so that *Tulip* sails slowly southwest. I move forward to the bow and dangle my legs over the rail, watching the remnants of the sunset, as seafarers have done for centuries.

Frank Bullen describes Abner Cushing, the melancholy Vermonter on board the whaler *Cachalot*, "sitting on the rail alone, steadfastly gazing down into the star-besprent waters beneath him, as if coveting their unruffled peace."[3]

Evening is the finest time of day, as the searing heat subsides, and the sky is briefly neither fierce white nor deep black. In the turmoil of this existence, each quiet spell alone is to be treasured. On our last visit to Trincomalee I had even reached the state where I welcomed a bureaucratic trip to grimy Colombo — just to be by myself for a few days.

Margo, too, needs her solitude. She is sitting against the mast ten feet behind me, gently cradling her guitar. She plays and sings softly. I hear snatches of Newfoundland ballads, songs she sang on Silver Bank, where live music was a regular element of our times at sea. In this heat and with so much stress, her music is a rare jewel.

As dinnertime approaches, Jonathan calls out the forehatch. "Margo, Hal, would you like a beer?"

Beer sounds good. Jonathan passes up two warm bottles — there is no refrigeration on *Tulip*. Sri Lankan beer is nothing special, especially when warm, but the bottles are large, and they provide some refreshment after a tiring day. However, this beer is very, very weak, and Margo calls out, "Oh, yuck! It tastes just like water."

"Mine, too," I add. "They must have made a mistake during the brewing."

"What about Sri Lankan Customs?" suggests Margo with a knowing grin.

"Oh, no! The crooks . . ." We had a problem getting this crate of beer past the customs post on the wharf at Trincomalee. The customs officers reserve the right to investigate anything we take on or off the boat and, if they are feeling obstructive, can make life difficult. As we were carrying the crate of beer past their post they had stopped us, asking for a bottle each as a bribe to let it through. I had had enough of bribery, and refused, so they seized the crate until we obtained an official chit allowing us to bring the beer on board. In the meantime, they apparently drank the beer, filled the bottles with water and replaced the caps.

We were already smarting from a similar incident in which they had stolen our best fuel jerrican. Although I sometimes grow angry with people or situations, I usually remain in control of my emotions. But Sri Lankan Customs seem to have the nearly unique ability to conjure up my pure rage; on the only few recent occasions that I have felt almost uncontrollably like hitting someone, it has always been when faced with a Sri Lankan Customs official.

"Oh, come on, Hal," says Margo, laughing. "See the funny side. Those guys could never afford a beer otherwise."

She's right; one bottle of beer probably represents more than two days' wages for the average customs official.

We have received much more from Sri Lankans than they have taken from us. Of all the kindly Sri Lankans we have known, even including one or two customs officials, I think especially of Thunga. Thunga Prema has become a particular friend of Gay's; he has been an enormous help to her research on the dolphin by-catch. He works for a fish buyer in the Trincomalee fish market and knows all the fishermen. A quiet, studious Sinhalese, Thunga has become fascinated by what Gay tells him of dolphins; he willingly keeps a record of the number caught each day in Trincomalee. His interest is such that Gay is now teaching him to identify the different species, so that the information he can collect will be even more valuable. Thunga has also been helpful in other ways. He has arranged for our transport onshore and has taken me across Trincomalee on the crossbar of his bicycle. He has even assisted our work offshore with the sperm whales, by explaining to the fishermen what we are trying to achieve and how the noise of their engines disrupts our recordings. Now we are rarely bothered by fishing boats whose crews are looking for cigarettes.

With the assistance of Thunga and a few fishermen in other Sri Lankan ports, Gay has begun to appreciate the extent of the

Sri Lankan dolphin by-catch problem. She estimates that at least forty thousand dolphins are being killed each year, and it is likely to be increasing. The significance of this figure depends greatly on the species that are affected, the size of their populations and their rate of reproduction. Gay has found that at least ten species are involved, but at present has no idea of their numbers or reproductive rates. It would take a whole new study specifically targeted at the dolphin by-catch problem even to obtain a partially satisfactory scientific grasp of the situation.

In the meantime, it is important to try to alleviate the by-catch. Small changes in fishing techniques might considerably reduce the number of dolphins caught, and experiments with different gear types and methods could show which directions to take. In the eastern tropical Pacific tuna seiners have substantially lowered their by-catch of dolphins by making small changes in the type of nets they use and by changing the methods by which the nets are hauled.

Dinner is finished and it is my watch. The others have turned below to sleep, and I am left to wash the dishes and track the whales. First I check the directional hydrophone. Yes, the whales are still close. Then I draw two buckets of seawater, find the detergent, which has been rolling around the cockpit floor (now devoid of rocks, which were dropped toward unsuspecting sperm whales, without obvious reaction), and holding a torch between my knees, I begin to wash and rinse pans, bowls and chopsticks, which most of us prefer to forks when at sea. I stack the dishes in the corner of the cockpit to drain dry, and hope that *Tulip* will not lurch too violently and scatter them. Every ten minutes I check the directional hydrophone — the clicks are still clear. The washing-up takes a little over half an hour, at the end of which the clicks have decreased to a grade three. It is time to be moving on after them.

I scramble forward to the bow, being careful not to tread upon Jonathan, who is stretched out on deck above the main cabin — the most pleasant place to sleep on *Tulip* in fine weather. I enjoy feeling the warm deck on my bare feet; I experience pleasure in just moving. We do so much standing and sitting, and have so little opportunity to walk even the length of *Tulip*. I like to do the sail work in the dark, to feel for the ropes and shackles and canvas. It is welcome exercise to heave on the winch handle to tension the sail, then to hurry back to the cockpit to trim the sheets and adjust Clarence until *Tulip* settles easily on her course.

I check the ocean ahead for fishing boats, fill in the data sheet and then lean back to watch the stars. In the days before we had a satellite navigator stars were beacons by which we found our position. Making sextant star sights every morning and evening I knew the names and patterns of the stars and constellations. But since the satellite navigator they have faded in my memory.

Fifteen minutes later it is ten o'clock — time to record. Through the headphones I hear the clicks of the whales sounding, sounding . . . . I take the headphones on deck, and beneath the stars, surrounded by the ocean, let myself sink into their realm. It is at night, alone on deck, that I feel closest to the whales. The resounding clicks and creaks and codas seem to tell of their struggles and joys. Those wrinkled, almost prehistoric bodies that we see in the daylight are hard to penetrate. They seem too monumental to be living, feeling animals. Alone on deck under a sky almost as deep black as the whales' world below, I listen to their clicks, and *Tulip* becomes an extension of my body. The sails are my clumsy flukes, the rudder becomes coarse flippers, as I join the whales in their wanderings.

Part III

Sri Lanka to the Maldive Islands
Autumn 1983

10

SRI LANKA

September 25 - October 16

TOWARD THE END OF OUR LAST SEASON, in April, Jonathan and I decided to split *Tulip*'s third and final field season into two parts, timed to coincide with the more favorable weather between monsoons — from October to December 1983 and from February to April 1984. We had never sailed off Sri Lanka in the autumn, although this was when the Yankee whalers had made their catches of sperm whales near Colombo during the 1850s. There were several issues we wished to investigate. Most significantly, we wondered whether the large male sperm whales would be off Sri Lanka, joining with the groups of females to mate. Sperm whales have a fifteen-month gestation period, so that our sightings of small calves in March suggested that the Northern Hemisphere autumn would be the time that the huge males migrate from the Antarctic to mate. Modern whalers had found them traveling north past the South African coast in October. Were any heading for Sri Lanka? There were other intriguing questions about Sri Lankan waters in the autumn: Would we see blue whales? And what other members

127

of Melville's "cetaceous tribe"? There were also more basic questions needing answers: we did not know how favorable the weather would be for research.

We also wished to continue the highly productive February to April study. I had commitments at home in Newfoundland, and Jonathan had a pile of data awaiting analysis in Cambridge, so we divided the studies. I would skipper *Tulip* from October to December 1983, and Jonathan would take over between January and April 1984. This would allow us each to structure the research according to our different inclinations: for Jonathan the creative, intuitive investigation and development of new techniques; for me the rigorous pursuit of statistically significant data, my scientific bread and butter.

The results of our research publicized by World Wildlife Fund, and particularly the underwater film sequences of lithe, twisting sperm whale bodies, had alerted many of those interested in benign whale research to the *Tulip* Project. Versions of the film have been shown at several meetings and conferences. Because of our unreliable engine, we decided not to take *Tulip* to the Seychelles for the publicity trip in May. In order to try to minimize the political fallout, I flew there myself to attend the conference and show the film. I also visited the Netherlands and Boston in order to publicize World Wildlife Fund, the whales and the *Tulip* Project. Although it was flattering to have our work praised and admired, I did not enjoy much of this peripatetic existence.

The kind of publicity that was given to the *Tulip* Project tends to raise the hackles of other scientists, who, rightly so, judge research by careful scientific papers, not by newspaper articles or films. The many months that are needed to analyze our data, and the year or more of lead time for scientific publication, meant that we had virtually nothing rigorous to show to the sticklers. In both Sri Lanka and the Netherlands we were criticized for not knowing what we were doing and, in particular, for ignorance of the literature on the whales of Sri Lanka.

In our opinion, these criticisms were not justified, but they hurt nonetheless.

Of far more concern than the publicity and the scientific criticism was the news that reached me in late July while studying finback whales among the cold, rugged bays of northern Newfoundland: violent racial riots had spread over Sri Lanka. The Buddhist Sinhalese majority, who consider themselves the true inhabitants of Sri Lanka, and the Hindu Tamil minority, originally from South India, had always had their differences. These were fanned by the British, who had imported Tamil laborers for their tea plantations and, as they had done throughout much of the British Empire, successfully pursued "divide and rule" policies.

During our previous seasons in Sri Lanka, we had occasionally heard of Tamil terrorists, the "Tigers," who were fighting for a Tamil homeland that they called Eelam, in northern and eastern Sri Lanka. They had attacked a few police stations near Jaffna, in the northern Tamil heartland. There seemed to be only a few Tigers, and their support in the Tamil community appeared limited.

But in July 1983, a series of incidents sparked an explosive situation. Acquaintances from the various ethnic groups have given us different variants of the chain of events, but this version from a member of the relatively uninvolved European "burgher" minority is probably as impartial as any: soldiers from the Sinhalese security forces had raped some Tamil schoolgirls in the Jaffna area. The Tamil Tigers retaliated by ambushing a Sinhalese army patrol and killing the soldiers. Roused by the soldiers' funeral, and possibly organized by shady political groupings, members of the Sinhalese population rose in anger throughout the island, killing Tamils and burning their property. The security forces, composed mainly of Sinhalese, either did little to stop the riots or actually took part in them. In Trincomalee, Sinhalese sailors of the Sri Lankan Navy rampaged through the streets, destroying and killing.

The terrible reports relayed by Canadian radio caused me considerable concern about the fate of our Sri Lankan friends and of *Tulip*, which had been laid up at the boatyard of Constellation Yachts in Trincomalee at the end of the last season. Most worryingly, Gay was in Sri Lanka at the time. I later learned she had been sheltered by Cedric Martenstyn during the violence, and had safely returned to the United States. *Tulip* had also escaped unharmed, but we were concerned about the effects of the riots and whether we would be able to do our research in the autumn.

I managed to reach Cedric by telephone from Cartwright in Labrador, a cold, isolated settlement on the edge of the Arctic. The long static-filled pauses that interspersed our conversation signified the separation (in more than just miles) of the world on the other end of the line.

"Everything is under control now," Cedric assured me. "You should not have any major problems doing your research, although things may take a little longer to achieve than usual."

As the northern lights played over the cold night, I walked back to the boat we were using, filled with thoughts of the troubled tropical island so very far away. Although reassured by Cedric's words, I wondered how the terrible violence would have infiltrated the fabric of Sri Lankan life. We might be able to work there, but would Sri Lanka be the same exuberant, friendly and colorful place we had known?

Gay had been uncertain whether she would accompany me in the autumn or Jonathan in the spring, or neither. The main consideration was her work at Yale, where she had been accepted as a graduate student. When she reached the United States in early August, she was able to make her decision: she would join Jonathan; as Martha and Margo had begun nursing and teaching careers respectively, I would need a completely new crew.

Three of *Tulip*'s new crew have sailed together with me off Newfoundland, but it is the fourth who seems most in place on board *Tulip*. Caroline Smythe is the identical twin of Martha Smythe. I first met her at the end of our winter study in April 1983, when she came out to Sri Lanka to spend the summer traveling in South Asia with her sister Martha. We talked one night in a small guest house in Trincomalee.

"I'd very much like to join your fall study. Is there any chance?" asked Caroline wistfully.

"I don't know. What sailing experience have you had?"

"Oh, Caroline has sailed more than me," chipped in Martha.

"Well, there were the two months on *Tiki* . . ." began Caroline.

"*Tiki*???" I exclaimed.

"Yes, Leaky *Tiki*, in the Caribbean — "

"Oh, no! You poor thing!" I cut her off. "I think there's a good chance we can take you." *Tiki* must have been the leakiest, most rotten and least seaworthy schooner in the West Indies. I had spent a week working on board her in 1980, but the pitiful state of what had once been a lovely boat, and the temperamental unpredictability of her ex-band-leader skipper/owner, meant that a week was about the average longevity of *Tiki*'s crew. To have spent two months on *Tiki* spoke eloquently of Caroline's courage, tolerance and ability to deal with unexpected situations — the very qualities that were so valuable on board *Tulip*. Like Martha, Caroline has an easy, pleasant manner, but she seems somewhat less exuberant and rather more scientifically directed than her sister. Dependable, thoughtful and at home on the sea, Caroline is one of the finest crew members *Tulip* has had.

But she is closely rivaled by Chris Converse. Brought up in and around boats of all descriptions in the area of New Bedford, Massachusetts, the largest nineteenth-century Yankee whaling port, Chris was U.S. National Champion in the International 10 Square Meter Sailing Canoe last year. A sailing canoe, which bears about as much resemblance to the more familiar paddled variety as a Formula One racing car does to a bicycle, is a racing dinghy taken to an extreme in the quest for speed. Athletic and wiry, with very curly blond hair, Chris has a degree in engineering, and plenty of experience in the building and repairing of boats and marine instruments. He is laid-back, cheerful and easily sees the humorous side of almost any situation.

The youngest crew member is Phil Gilligan, a zoology student at King's College, London University. Phil is big and strong, and would appear intimidatingly tough if he did not have such a merry, warm smile. A no-nonsense realist with military inclinations and very short hair, Phil was an obvious butt for the "liberated," left-wing idealists who formed the remainder of my Newfoundland crew. They teased him at every stage, even hid his wrist exerciser, but throughout Phil remained cheerful and amused, giving back as well as he received. He is keen to learn about every phase of our work, and willing to undertake any task, no matter how unfamiliar. He was more than happy to interrupt his university education to join us in the Indian Ocean.

Finally there is Lindy Weilgart, to whom I am closest, and who is therefore hardest to describe. Tall, with long dark hair, Lindy had been brought up in Iowa far from the ocean. Notable for her firmly expressed opinions, she is as easily outraged by greed and waste as she is overwhelmed by the trust and innocence of a young animal. Rather unreasonably for an Iowan, Lindy developed a passion for whales. This was finally realized when she became a graduate student at the Memorial University of Newfoundland, where she studies the sounds of Atlantic pilot whales (*Globicephala melas*) and where we met.

We each have different responsibilities, although there is much help and overlap. Chris looks after the boat, sails and rig, Caroline is in charge of food and supplies; while Lindy is the fiercely zealous accountant, guarding W.W.F.'s every penny. Phil is continuing Gay's work on the dolphin by-catch, while I try to deal with the bureaucracy and keep the engine running. At sea we each have additional scientific responsibilities. But, as in the preliminaries to each previous season of the *Tulip* Project, it often seemed we would never leave port.

For the three weeks since we arrived in Sri Lanka we have had to struggle against both expected and unexpected obstacles. The most alarming was the first: as our airplane touched down at Colombo Airport, Chris was doubled over in pain and had turned very pale. Through Cedric Martenstyn we found a doctor, who diagnosed appendicitis. Chris was rushed to a hospital and operated on. His appendix had burst.

The hospital was small but clean and spacious, set around a lovely garden. Chris received excellent care, and after he'd endured two days of great pain, his health began to improve rapidly. Our roles were now reversed, and as we toiled through the heat and hassles of Colombo, it became the highlight of the day for us to visit Chris in his peaceful paradise. Dirty and disillusioned, we would struggle in for the evening visiting hours. Chris would get up from his bed and greet his visitors. "There, there, now. Just put all your things down, lie down on the bed, relax and tell me all about it."

He would listen patiently to all our problems and dispense his wry wisdom. However, hardly having seen the real Colombo, he found it difficult to empathize with our difficulties. He soon became frustrated by his inactivity. "Phil, you've got to bust me outta this joint!"

"You're crazy, Chris," replied Phil. "It is *desperate* out there. The trunk that contained our most valuable equipment is lost somewhere between here and London, Caroline has been robbed of her money twice on the same day, Lindy and Hal have had their cameras stolen, Sri Lankan Customs wants our money and N.A.R.A. has demanded everything else. You have the best deal going."

We had been shuttling around Colombo, trying to clear airfreight through customs, make arrangements for the film crew, who wished to return for more footage, satisfy the increasingly stringent bureaucratic whims of N.A.R.A., find our lost trunk, buy new cameras and . . . A few of the officials we called on were helpful and some were hostile, but almost none could see us immediately. We waited in more offices than I wish to remember and were offered many cups of "tea." Although producing some of the world's finest teas, Sri Lankans themselves, in the words of a Sri Lankan friend, "use tea purely as a medium for swallowing dissolved sugar." Despite the general unpalatability of the liquid that was placed before us, those cups of tea were important symbols of the hospitality of Sri Lankans.

During our bureaucratic peregrinations we used most of the forms of transport that Colombo has to offer. The Morris Minors, which form the bulk of the Colombo taxi fleet, disappeared from English streets twenty years ago, but here they were kept running year after year, thanks to extraordinary resourcefulness. However, we sometimes found ourselves having to push the taxi in which we were paying to ride. Cheaper than taxis and usually more reliable were the three-wheelers, with one wheel at the front and two in the rear. The driver steers with handlebars, and one or two passengers occupy a seat behind him (although we once fit in four). The occupants are protected from sun and rain by a canvas hood. A plastic Buddha (or a Jesus or Hindu goddess) often hangs from the windscreen, proclaiming the driver's religion. Cheapest of all are the ramshackle government buses and competing "private" Japanese-made minibuses, especially good for traveling into the center of Colombo, the district called "Fort." As a minibus approaches a bus stop, the conductor, usually a small boy, hangs out the door, screaming something like "Pettahfortbambalapityakolu-pityadehiwelawelawattamounlaviniamoratuwa."

If I believed that I heard "Fort" somewhere in his patter (a slurring of the names of the communities on the bus route), I would climb aboard. Near the start of its route, the minibus was usually fairly empty; there might be one spare seat. But as it made its way into central Colombo, more and more people were taken on. Even when it seemed as though the interior of the bus was a solid mass of humanity, there was room for ten more. All passengers would be in simultaneous contact with an amazing variety of parts of other bodies, both male and female.

The minibuses sped erratically through the multiformity of Colombo, past wealthy suburbs with huge trees, blossoms and wide clean avenues, through busy commercial districts and rows of middle-class homes, neat and orderly bar the occasional charred ruin where Tamils used to live. We might pass a squalid dump, skirted by a jumble of hovels. The passages between the houses, each little more than a few palm fronds, were churned to a muddy swamp by bare feet and rain. People could be seen picking over the garbage, searching for a meal.

On reaching Fort, in central Colombo, the bus would quickly empty. By now I know this area and its sights. There is the hawker of glassware, who spends his day rhythmically banging a tumbler against a wooden block to show its strength, the dwarf beggar, whose legs end at the knees. With wooden pads strapped to his elbows, dressed in clean shorts and shirt, he crawls around his tiny patch of central Colombo, gazing intently upward at potential donors. Is this life in the raw? Or is nothing quite what it seems? Are the beggars really wealthy enough to set up money-lending businesses, as the newspapers say? Are the happy, laughing youths the uncontrollable mob that burned and murdered two months ago?

Everywhere we went there were signs of violence. Most visible were the rows of burned-out houses and shops that had belonged to Tamils, Hindu temples that had been converted into refugee camps, patrolling armed soldiers, and throngs of Tamils at the airport, waiting for a flight to India. The Krishna Palace, the tiny Tamil restaurant where we ate so often last January, was a blackened shell. The Tamil owner of the international telephone and telex office, where we had waited when trying to call our loved ones at home, had been murdered, and his family had fled.

However, these were only the surface wounds. Throughout all of the Sri Lanka that we knew, people were stunned by the summer's violence. The Tamils mourned their relatives who had been killed, and grieved over their burned houses and lost

135

possessions. Sheltering in refugee camps or in the homes of non-Tamil friends, they pondered an uncertain future. Should they rebuild their property, which might be razed again at the whim of the Sinhalese mob; move to the almost entirely Tamil, but overcrowded, northern part of Sri Lanka around Jaffna, where they would be unlikely to find jobs; or immigrate back to India, from where their ancestors had migrated, often centuries before?

The Sinhalese, although perplexed by what had happened, seemed remarkably unrepentant. Those whom I talked to seemed to feel that, although they regretted the violence, the Tamils almost had it coming to them. There seemed little foundation on which to rebuild a nation.

Superimposed on personal tragedies and communal tension was a feeling of fear and lawlessness, symbolized by the charred buildings. The security forces had done little to quell the riots. In some cases army and navy personnel had encouraged the violence, in others they took part in it. Crime was rampant. On several occasions during our two weeks in Colombo we had money and possessions stolen, something that had never happened during the two previous seasons.

It was some relief to move to the more peaceful surroundings of Trincomalee, although here, too, the landscape was scarred and the people afraid. As in Colombo, Tamil houses and shops were charred ruins, and along the peaceful Sri Lankan lanes were checkpoints manned by newly armed and very nervous soldiers.

Constellation Yachts had done a fine job of refitting *Tulip*, but because of delays in clearing parts through customs, it took a few more days for the engine to be reassembled and mounted. During this time we were privileged to be guests at the Hindu festival of the goddess Sarasvati, patroness of work and the workplace. The main workshop at Constellation Yachts was gaily decorated with plants, flowers and portraits of the goddess. There were songs, prayers, food and Hindu music. We and all the workers, lathes, power drills and other machines received elaborate Tika marks on our foreheads (or its closest approximation) — the third eye of the Hindu. The ceremony emphasized the Hindu belief that all things animate or inanimate, from priests to birds to power drills, are related.

Eventually *Tulip*'s engine was mounted, her bottom painted, and she was ready for launching. She was placed in a steel cradle on a trolley at the top of a marine railway running down into the water. The cable holding the trolley at the upper end of the slip was released, and she slid down the ways toward the sea. But before floating away, *Tulip* suddenly stopped. Her waterline was still a foot and a half above the water surface, and the trolley had reached the end of the railway. We asked the Constellation workers how they had hauled her out. "Oh, we took her out on Poya Day!" Poya Day is the Buddhist festival of the full moon. It is also the time of spring tides, when the sea rises to its highest level. The typically Sri Lankan solution of the Constellation workers to our problem was "Wait till the next Poya Day," which was only one week away. But our time was fast disappearing, and I was desperate for the sea. Chris and Phil came to the rescue. They rented ramshackle diving equipment and, with the help of some Constellation workers, used cold chisels and hydraulic jacks to dismantle the welded steel cradle around *Tulip*. It took a day of much effort, with Chris's ingenuity directing Phil's strength. Finally the last weld in the cradle was snapped by a hefty jerk from a line attached to a speeding powerboat, and *Tulip* fell unceremoniously into the water on her side. We were released from the land.

The sea, the sea. She cradles *Tulip*, rustling where parted by the firm stem. Sometimes she heaves and brushes my toes, which hang from the bow — a gentle welcoming caress, but holding, far beneath, the unknowns that we seek. Her moods are the warp through which this life is woven: the immense roar of the storm wave, the eerie oily deep calm stretching the present, the boisterous good nature of a fresh breeze. Tonight she is flecked with sparks of phosphorescent plankton, the light of life.

At this reunion, the sea is a sanctuary. A serene ocean around a troubled land is as welcome as a safe port in a storm.

The sea is a little greener and murkier, the sky emblazoned with thunderstorms and the air pleasantly cool, but it is the same sea I left six months ago. Earlier today, as we left Trincomalee, we made a standard thirty-nautical-mile transect due east from a prominent landmark in Trincomalee called Swami Rock, watching and listening for whales. Swami Rock has a gaudily painted Hindu temple, a lovers' leap, tame monkeys and a fine view of the area where the blue whales feed. We had made these Swami Rock transects every two to three weeks through the last season. We use them to keep track of the changing distributions of the whales off Trincomalee.

Today we saw one blue whale a few miles from Swami Rock, and later, when we reached deeper water, there were the familiar clicks of sperms though the hydrophones. Tomorrow we will try to follow them.

11

EAST OF SRI LANKA

October 17 - 27

IN THE COCKPIT STREWN WITH ROPES, Chris uses all his wiles to keep our spinnaker Wuff-Wuff round and drawing. As the gentle breeze shifts almost imperceptibly, he trims a few inches on the sheet, eases the guy a touch, bears off several degrees, to keep the spinnaker full and pulling *Tulip* north toward Trincomalee. Like the conductor of an opera, Chris coaxes the temperamental diva of a spinnaker to carry us, adjusting the heading and setting of the sails to its whims.

Last night a little more wind possessed this sensitive sail, giving it uncontrollably destructive power. All Phil's strength at the helm was inadequate as a squall filled Wuff-Wuff, flinging *Tulip* onto her side, in what is called a "broach." We quickly doused the unruly spinnaker, relieved that the port spreader, which has developed an alarming crack, had not parted and brought the mast down.

An adverse current has slowed our progress, and empty fuel tanks leave us dependent on the sails. But much as we look forward to freshwater showers and cool drinks, we have no great desire to return to land. During these past ten days the sea has filled our lives so vividly and so powerfully that we welcome the opportunity that this gentle breeze gives us to relax before we reach the confused urgency of the land.

"Hey, Caroline," Chris calls out, "there's a bunch of birds ahead." Caroline, our seabird specialist, grabs binoculars and her bird identification book and is climbing on deck, when Chris adds, "Phil, I think there are dolphins underneath the birds." Phil takes his camera out of its case and follows Caroline. Five hundred yards ahead a patch of frantic activity stars the smooth ocean. The birds, probably terns, are hovering and diving. Are those dark shapes scything the turbulent waters the fins of dolphins?

We approach to within a hundred yards, but the feeding frenzy is dynamic, and not easy to track. The dolphins, mostly spinners but with a few striped dolphins interspersed, burst from the water unpredictably, often preceded by small tuna. Are these tuna their prey, or are they common predators on a smaller food source? The terns hover a few feet above the water, their dark caps and long beaks pointed downward, watching intently until they see potential food; then they plunge. On the surface swim a few shearwaters, scrabbling among themselves, occasionally making short dives for food. It seems probable that the dolphins have driven the prey to the surface, making it available for birds. Many of the birds may have found the food source by following the dolphins.

The system of life in the oceans is more complicated than a simple list of who eats whom. Species may assist or hinder one another in relatively complex ways. One kind of predator may change the behavior of a prey species to the benefit or detriment of another. This prey school may have been driven to the surface by the dolphins, and thus become available to the relatively shallow-diving birds. Small fish may school tightly for protection when attacked by predatory larger fish, but this makes them ideal food for bulk-feeding baleen whales. Similarly, a prey species may distract, confuse or satiate a predator to the benefit of its other prey. A well-fed shark is much less dangerous than one with an empty belly.

It is very hard to give a reasonable assessment of the true importance of any species, or group of species, in the marine ecosystem. To begin with, there is no clear consensus as to what we mean by "important." Is pure weight, the "biomass" of a species, a measure of its importance? If this is our criterion, then whales, which for a given ocean area have similar biomasses to sets of much smaller animals or even plants, are very important.[1] However, if we are more concerned with productivity than with biomass, then whales are clearly outperformed by the plankton, which grow and reproduce more quickly.

The sperm whale does seem to have a crucial role in the marine ecosystem. It has been calculated that, before it was reduced by whaling, the sperm whale population ate more squid each year than today's entire catch of all fish species.[2] Thus, in terms of tonnage consumed, sperm whales once had a more significant impact on the life of the ocean than the total human fishery.

Sperm whales mostly eat squid, which form only a small part of the human fishery. Scientists speculate that if sperm whale populations are decimated, then the squid that they eat might multiply, and consume more valuable commercial fish. Thus the sperm whale restricts the numbers of a potential competitor to man for the harvestable wealth of the ocean.

Sperm whale feces may also be important to the health of the oceans. Plants can only grow in the upper waters, where sunlight can provide the energy for life. When plants die, or animals that have eaten them defecate or die, the nutrients (especially phosphates and nitrates) that are essential to the process of life sink downward into the abyss. This rain of nourishment provides sustenance for the creatures of the depths, including, ultimately, the sperm whale. But the all-important phosphates and nitrates are now out of reach of the plants at the surface, the base for all marine life. It is vital that these nutrients return to the plants in the upper sunlit layers of the ocean. Sometimes this happens when deep currents meet land or join together, rising to the surface to form "upwellings" where life can proliferate. But the sperm whale, by eating creatures of the deep waters and defecating their remains at the surface, is also turning the wheel of existence.

The surface of the ocean, which during the spring was a pure barren blue, is now tinted with the green of growing plankton. Dolphins, birds and turtles, which in April seemed virtually prisoners of the narrow band of sea along the edge of the continental shelf, where the nutrient-rich waters upwelled, are now spread over deeper waters. The shelf break is still apparent, with overfalls (where the waters churn on even the calmest days) and more than its share of the surface life, but the contrast between the desert of the deep ocean and the oasis of the shelf break has been blurred by the churning of the southwest monsoon. Ashore, the monsoon gives rain so that animals that had huddled around water courses may disperse; at sea it is the wind that stirs the production of life, making previously desolate regions attractive to birds or dolphins.

Several hundred feet down the ocean seems more constant. Although presumably affected by the rates at which dead plants or animals or their products fall from the sun-blessed surface, the abyss seems to maintain its own slow but steady pace. And sperm whales, rooted in the ecosystem of the deep waters, can lead a more ordered existence than the large baleen whales. Blue whales, and their relatives, must voraciously capitalize on transitory plankton "blooms" or concentrations of spawning fish to build up their blubber layers for long fasts when no suitable food is available.

On our second day at sea, we found a large group of sperm whales. Despite the new crew's excitement, among whom only Phil had seen sperm whales before, we tried to systematize our research techniques. We had left onshore most of the equipment that the previous seasons had shown to be only marginally effective, such as video and movie cameras, but it was soon apparent that of the present crew only Caroline could manage the complex data sheets that her sister Martha had handled during the spring. Lindy and I tried to devise a simpler system for recording data. This project, the demands of training a totally new crew everything to do with following sperm whales and the usual round of engine problems left me feeling close to my limit. But I was happy. We were at sea and with the whales.

Using our directional hydrophone, we managed to track this first group of whales, called Watch Number One, all day, as they moved steadily southward parallel to the Sri Lankan coast. In the evening a dramatic thunderstorm swept out from the island, disrupting life on *Tulip* with two hours of pyrotechnics, an artillery barrage and a deluge. Evening thunderstorms seem to be a regular feature at this time of year. However, we followed the whales through the storm until two o'clock the following morning, when the power supply to our tape recorder failed. In the two hours that we needed to repair it, we lost the sounds of the whales.

Twelve hours later we had picked up a second group, Watch Number Two, which contained at least nineteen adults and one calf. The calf seemed surprisingly small, about thirteen feet long. From our observations in the spring, we had asssumed that most calves were being born around March, but this young whale seemed too small to be seven months old. A day later we lost the clicks of this second group. We had made a reasonable start, but it was frustrating; each loss of contact occurred just as we seemed to be beginning to recognize the individuals in the group and to learn something of their behavioral patterns.

For a day and a half we sailed north past Trincomalee, but heard no clicks. This was a generally welcome pause. We could rest and plan the development of the research. The crew were coming together, learning their roles and the challenges and joys of life with the whales.

At 8:00 a.m. on October 21, we heard the loud cacophony

of clicks made by a large group of sperm whales. The weather was calm; we were rested and ready. *Tulip* was soon up with the group of whales, which contained a really tiny calf (less than thirteen feet long). It seemed curious and often approached *Tulip*. Chris noticed that the calf still had its umbilical cord protruding. It cannot have been more than a few days old. The calf was almost always accompanied by at least two adults. They pushed it around with their foreheads, and gently touched it and sometimes each other with their flippers. We all had an opportunity to swim behind *Tulip*, and the whales revealed themselves.

Like other whalers, Herman Melville had watched the backs of sperm whales brushing his above-water existence, marveled at the hurtling mass of a breaching animal and stared at bloated carcasses tied alongside whale ships. But these views of sperm whales were frustratingly incomplete. Melville concluded "there is no earthly way of finding out precisely what the whale really looks like."[3]

There may be "no earthly way" of perceiving the shape of the whale, but the sea has a way. With our eyes beneath the waves, face to face, the sperm whale reveals himself.

I am not a good swimmer. I normally feel the need of a boat through which to interpret the ocean. Without whales, I swim slowly and briefly, soon returning to the floating home that I know. But when I am with the whales their spirit takes mine; the boat is nowhere; the normally cumbersome business of breathing through a snorkel is automatic and forgotten. Without being asked, my legs propel me steadily forward toward the huge magnetic shapes. The vast forms undulate — dark gray liquid images pouring through the sea.

Now three bulbous heads grow into my field of vision. The smooth case, the "brow," which surrounds the chamber of oil and enormous brain, entrances me, as it did Ishmael: "This high and mighty god-like dignity inherent in the brow is so immensely amplified, that gazing on it, in that full front view, you feel the Deity and the dread powers more forcibly than in beholding any other object in living nature."[4]

Clicks thud through my body as two adults with a calf between them swim slowly toward me and I, spellbound, toward them. Their aspect slowly transforms — they pass just beneath. The closest whale is six feet below me. Its smooth case and wrinkled body flow on. Just beyond, the calf's eye follows me as they pass. The bodies rock slightly, transmitting a powerful force to the water through the flukes. The forms melt away. In vain I try to swim after them, but the spell is broken. I am once again a human floundering in an alien element, waiting for *Tulip* to return.

Toward evening the whales split into clusters of between one and three animals, raised their flukes and dived far beyond our reach.

Through the night we followed the clicks of the group. In turn each crew member would have the "watch" — the responsibility of listening for the whales, filling in the data sheets and maneuvering *Tulip* to maintain contact. The others could rest, but few of us slept well after all the excitement.

At 8:15 a.m. I joined Caroline and Chris on deck.

"By the sounds of the clicks, the whales are close, but I haven't seen any yet," reported Chris.

"Would you like a cup of tea?" asked Caroline. "The kettle's just boiling." She went below and came up with three mugs of tea, and we sat around the cockpit, watching the horizon for blows. It was a still morning, with only slight swells from the east. After a few minutes we saw some faint puffs about four hundred yards away.

"Well, I suppose we'd better get over there," said Chris, who was officially on watch. I went below and started the engine. But as Chris put *Tulip* into gear, Caroline called out, "Wait a moment. There's a whale just off the port beam. It looks as though it has a callus — probably a female." Chris slipped the engine into neutral as I went below to get a camera. We were in no particular hurry. The day had just begun, and we were still satiated with the previous day's excitement.

As I returned on deck, the whale began to make unusual movements, flexing at the middle to show first both fluke tips and head, with her body U-shaped, and then arching her back. A few seconds later she repeated these contortions. We called down the hatch, "Hey, Phil, Lindy. Come up here. There's something unusual going on."

The whale then rolled on her side, with her belly toward *Tulip*. A rush of dark blood and a solid object were expelled from the genital area right at the water surface.

"Did you see what I think I saw?" gasped Chris.

"There it is, just beside the mother's head," exclaimed Caroline, pointing at a twelve-foot-long gray object. It was a tiny sperm whale with wrinkled skin, curled flukes and a dorsal fin folded over. A whale had given birth right beside *Tulip*.

"I'll go up the mast to get a better look. Whose turn is it to swim with the whales?" I asked.

"Mine, mine," cried Lindy, eagerly dragging mask and snorkel out of the hatch. On reaching the mast position two minutes later, I saw the tiny newborn bobbing beside its mother. The mother expelled some more blood, but otherwise the two animals lay quietly at the surface.

However, their repose was broken after about two minutes by the arrival of another sperm whale, and then another and another. For fifteen minutes the newborn was surrounded by energetic elders. They thrusted and cavorted, apparently attempting to touch this new member of their group. The calf was rolled, pushed along by foreheads and once squeezed almost out of the water by the jostling adults. We saw flukes thrashing out of the water, spyhops and, once, an open jaw. What was going on? Why were the other adults so absorbed in the tiny calf that they almost seemed to be attacking it?

Twenty-five minutes after the birth, mother and calf were finally left in peace, although some dolphins circled about one hundred fifty yards away. However, the little calf seemed anxious for more action. It left its mother and started moving directly toward Lindy, who was hanging on to the rope behind *Tulip* twenty yards away. The tiny whale moved slowly, seeming to have little strength to power its curled flukes. As it approached Lindy, the mother glided over and nudged it away. Lindy could see the calf's umbilical cord, as well as the afterbirth protruding from the mother. The calf then dived under the mother in what could have been an attempt to suckle, but it soon returned to the surface.

The calf seemed intrigued with the small, pale object that hung in the water about twenty-five yards away — Lindy. Once again it swam closer. This time the mother did not intervene; instead she lingered twenty yards behind. Within touching distance of Lindy, the calf stopped, and for a few moments they watched each other before the calf returned to its mother. Like Ishmael, Lindy saw that "the delicate side-fins, and the palms of his flukes, still freshly retained the plaited crumpled appearance of a baby's ears newly arrived from foreign parts."[5] But Lindy was particularly taken by the calf's "*beautiful* dark blue eyes." Perhaps the calf was similarly impressed by hers, for two minutes later, it returned to Lindy. Once again their eyes met. But this time, instead of turning back to its mother, the calf moved on to *Tulip* and nestled against the port side of the boat, twisting around the leading edge of the keel. This is the area of the depth sounder transducer, which had been left on and was emitting steady pings. These might have sounded similar to its mother's clicks. Lindy, watching from behind, was worried that the boat might run it over, but *Tulip* was moving very slowly. The calf had one more close encounter with Lindy and then swam away in an apparently random zigzag.

As Lindy climbed back on board, Caroline greeted her. "That calf has spent more time with you than with its own mother!"

"Oh, *my baby!*" exclaimed Lindy, all her maternal feelings aroused.

Up the mast, I was worried. The real mother was nowhere to be seen. Perhaps she had deserted the calf because of its association with *Tulip*. The little whale looked lonely and pathetic.

It attempted a tiny lobtail, but hardly had the strength to lift its flukes clear of the water. What had been one of the most wonderful experiences of our lives had turned into a bitter lesson not to meddle with nature. We pictured the calf a starving orphan on the wide sea — all because we had watched its birth.

The calf finally straightened its track and set off slowly southeast. Phil checked the directional hydrophone and found that this was the direction of the clicks from the other whales in the group. At least the calf knew which way to go. We did not follow, as we had no desire to exacerbate what was already a very worrying situation. As the lonely-looking calf flopped ineffectually out of our visual range, our hearts were heavy. We sailed *Tulip* off toward distant blows from other members of the same group, and we hoped.

It was an enormous relief when, twenty minutes later, we saw the baby whale reunited with its mother. We approached close enough to check that it was indeed the animal with the scratches on its head that we were already beginning to call "Lindy's baby," and not the small calf we had seen the day before. We then left it alone to consolidate its relationship with its real mother.

The birth has given us a powerful psychological boost: we must be truly fortunate to be presented with such an experience. It also provided a scientific windfall. Most research results on the behavior of whales come after months, and usually years, of painstaking work. We had been given a significant insight in one hour.

There had been some previous observations of apparently newborn sperm whales by whalers and scientists working with the whaling industry, but no one had actually observed the moment of birth, or swum with a newborn. The birth of "Lindy's baby" had several interesting aspects.

First is the time of year — October. Sperm whales in the southern Indian Ocean are thought to mate in October and

November, and give birth between January and April. Both the mother of "Lindy's baby" and the mother of the other newborn in this group were six months out of phase. Is there, as we suspect with blue and humpback whales, a separate northern Indian Ocean population? And where did the father come from — the Antarctic or the Arctic? The small calf that we had seen in the previous group suggests that births at this time of year off Sri Lanka are not aberrant.

The mother's roll as the birth took place was particularly interesting. This meant that following the birth, the calf was immediately at the surface, ready to take its first breath, and did not need to be nudged or lifted up by the mother or an adult "auntie," as has been observed in some dolphin species. It is also interesting that there were, as far as we could tell, no other adults close to the mother at the time of the birth; she had no assistant.

However, the newborn calf seemed of great interest to the other adult members of its group. What were they doing by vigorously jostling the baby? It seemed no way to treat an animal only a few minutes old. Were they trying to shock its systems into normal functioning, much as we slap a newborn baby on the back to induce breathing? Although it seems unlikely to us, it is just possible that the adults were acting aggressively and were not pleased with this new addition to their group.

Most probably, calf and adults were engaged in a vigorous introduction. The calf will be spending at least the next five years in close contact with these adults, and they may have to baby-sit for it, feed it, or teach it. The calf will need to "imprint" on these animals almost as much as on its mother, and they will need to "bond" to it. The apparently rough treatment we observed may have been a most important phase of its life — an initiation into its social surroundings.

The wonderful fearlessness of the calf toward Lindy and *Tulip* is echoed in the report of some scientists working with the whaling industry off South Africa. A calf, who was probably less than an hour old, blundered into the whale catcher from which the scientists were watching. They assumed it was confused by their ship and its sonar.[6] I keep worrying that the early contact with people and boats may have harmed "Lindy's baby's" relationship with its mother.

We continued to follow the group of whales, listening for their clicks and watching for their blows. As usual, they congregated at the surface in the early afternoon, later splitting up to feed as the day wore on.

When night fell, the group was moving steadily southward parallel to the coast. I was terribly tired, but the crew had learned fast and managed to stay close to the whales throughout the night, without using the engine — always a satisfying achievement. The crew had also started recording more than the wind direction in the ship's log. On previous studies I had found this to be a good sign of a happy crew.

"2200. Lots of whales around in the dark." [Phil]

"2300. Zigzag across ze ohsion." [Chris]

"0200. Whales all around, and I mean everywhere!" [Caroline]

The third day with this same group of whales was also productive, but less intense than its spectacular predecessors. We sailed from cluster to cluster, taking fin and fluke identification photographs, recording headings, speeds, ranges and groupings. These data will allow us, months later in the laboratory, to describe the detailed social interactions within this large group.

Each hour we made a five-minute recording of the whales' sounds through the hydrophone. There already seemed to be a pattern: at night we heard the concentrated rhythmic clicking of many whales hunting in the deeps; during the day, and especially in the early afternoon when the whales were congregating at the surface, clicks were fewer, but we often heard the stereotyped coda patterns, which seem to have a more social function. There was a particularly heavy burst of codas at the time of the birth.

The diurnal rhythms of sperm whales can probably be traced to daily migrations of the squid on which they feed. Many species of squid, as well as many other ocean animals, move upward at night, presumably reacting to changing light intensity. This kind of migration might make them easier prey for sperm whales during the night, leaving the middle of the day as the most suitable time for resting and socializing.

We had several opportunities to try out a rotating depth sounder transducer that Chris had built, and after a few trials managed some long tracks (over five minutes) of the diving sperms. This is a much simpler system than the BBN-ing of last season, and seems more effective. BBN-ing required two or three towed hydrophones to be monitored continuously on an oscilloscope, and exact coordination with the helmsman, oscilloscope watcher and sail handlers.

On the third day with this group a long swell began to roll in from the south. This made our work, and especially swimming with the whales, difficult. However, Phil managed to swim with the calf of the first day, which now appeared quite nimble. It came up to *Tulip*'s hull beside the depth sounder, much as "Lindy's baby" had done the day before, but we did not see the newborn. That night Lindy wrote in the log:

"2300. Spermies clicking away. I'm missing my *baby*."

"2400. Wind's apickin' up. Looks like a big storm coming."

As the huge thunderstorm came over and the wind howled in the rigging, I worried about the newly discovered crack in the bracket holding the spreaders, which in turn keep the mast vertical. This crack may have been a result of my too-enthusiastic bouncing while watching the newborn whale from the mast position. We continued south with the clicks of the whales.

We had spent three full days following the same group. The wind kept up all the next day, making our work hard. We sighted the odd blow over the tops of the waves and took a couple of identification photographs, but mostly we stuck to the directional hydrophone and our hourly tape recordings.

As usual during rough weather, there were problems with the boat and our equipment. The magnetic tape became knotted in the tape recorder, and the mainsail developed a large rent. The former was fixed after some delicate fiddling; the latter was jury-rigged until we got to port.

The day's highlight was a sighting of "Lindy's baby." She was thrilled: "My *baby*'s back! Hi, cuteskie!"

The tiny calf's flukes and dorsal fin had uncurled. It had gained strength and now was able to swim steadily. And it had had to swim, for during the first two days of its life the calf had traveled about ninety miles.

That night the wind thankfully lightened somewhat, but a heavy swell was rolling in from the south. We were determined to stay with this group for as long as possible. None of us wanted to lose them, particularly not on our own watch.

A tern flew on board. Lindy named it "Budgie" and Chris warned, "Just so it's perfectly clear that that bird is going to have to stand watch, like any other crew member." Budgie provided some company for the watch keepers:

"0400. I'm teaching tern how to do watches and take data — welcome crew member number six to the *Tulip!*" [Lindy]

"0700. Budgie has begun to pass comment on the situation." [Phil]

At noon on the fifth day, the whales broke their steady southward progress parallel to the Sri Lankan coast and headed southeast, directly away from land toward the open expanse of the Indian Ocean. We had been at sea for ten days. We were low on fuel and food, and we were concerned about the damaged spreader. That evening, with great reluctance, we left "Lindy's baby" and its companions to sail back to Trincomalee. The whales had traveled one hundred eighty miles in four and a half days. Somewhere out there they are still traveling, clicking and touching.

12

SRI LANKA TO THE MALDIVE ISLANDS

October 28 - December 4

OUR SECOND TWO-WEEK TRIP TO SEA WAS a compendium of success and struggle. We left Trincomalee on November 3. Following *Tulip* out of the harbor was Flip Nicklin, the *National Geographic* underwater photographer, and his brother Terry, on board a local fishing boat they had chartered.

Soon after finishing our standard transect out of Trincomalee, Flip sighted blows. Both boats slowly closed in. A medium-size sperm whale lay at the surface. Chris, up the mast, noticed it had something wrapped around its flukes. We approached and saw a sixty-foot-long plume of fishing net trailing behind the whale. It must have become entangled, like so many dolphins, in a fisherman's net, but, unlike the dolphins, it had been strong enough to break through. However, the trailing net would gravely hamper the whale's attempts to swim and seriously disrupt its diving and feeding. Flip and Terry came alongside.

"Do you want to have a go at freeing it?" I asked.

"Sure, we'll try," they replied.

As experienced divers, they were aware of the dangers. To be trapped in the net would probably be fatal. Even a fairly shallow sperm whale dive would be too long, and encounter pressures too great, for a human. Flip, Terry and Phil, the most experienced swimmer of *Tulip*'s crew, put on flippers, masks and snorkels and swam over toward the whale and its trailing net. While Phil and Terry sliced away at the net with their diving knives, Flip photographed them. Unfortunately, before the job could be completed, the whale dived, and we never saw it again.

"The net had cut into the whale's tail-stock — it was open and raw. She seemed to flinch whenever the net tugged," reported Phil. I hope that the whale managed to free itself; otherwise it may have died a death more drawn out and even more painful than the entrapped dolphins.

On the second day out, we picked up a group of sperm whales off Trincomalee. They were moving slowly northward, against the current, which, during our spell in port, had strengthened to about four knots. Because of this, the whales were hard to follow. The increasing current meant that Trincomalee did not appear a promising place for sperm whale research in the immediate future. So we took Flip on board *Tulip* and headed southward around Sri Lanka to the area southwest of the island where the Yankee whalers had made their catches. We hoped the currents would be weaker there.

We found Flip a pleasant, easygoing companion. He has the strong and powerful frame that would be expected from such an accomplished diver, but he is also most gracious, a true gentleman. Flip gave us many hints about whale photography from both above and below the surface. As he talked of his experiences with humpbacks off Hawaii, orcas around Vancouver Island and other whales and fish in many watery corners of the world, I was struck by the utterly different natures of our work, even though we are focused on the same animals.

Whereas the most obvious quality needed for our kind of whale science is endurance — the ability to follow whales for days, to stay awake during the long night watches, to keep the standard collection of data rolling in and not to become distracted by the problems of port — most of Flip's working life is spent waiting. He may "stand by" patiently for weeks or months for that one absolutely unique opportunity to take a brilliantly revealing photograph. How well he does it will depend on his composure, intuition and strength for those few moments.

Flip coped well with the unusual discomforts of life on *Tulip*, and he was with us during some particularly trying times. As we rounded the Basses reefs, the southeastern corner of Sri Lanka, the wind rose to a full gale from directly ahead. We beat into it under storm jib and fully reefed mainsail. The cabin was the usual fetid, chaotic hellhole, made worse by some rotting mangoes and pineapples that had fallen from their hammocks in the saloon and had spent several days festering unnoticed, wedged deep among our equipment. Then the head (marine toilet) clogged. As we slammed to windward through heavy seas, I dismantled it, sitting on the cabin floor, washed by sloshing feculent bilge water, my body running with sweat in the steamy atmosphere. It was a task so completely revolting that it almost took on a perfection of its own.

After two days of trial, the wind moderated, and we found sperm whales. During the next week we followed various groups of sperms in the waters southwest of Sri Lanka. The weather varied from fair to just bearable. Unlike the first three groups we had tracked this season, who all seemed dedicated to a regular southward course, these animals meandered with no obvious goal or direction.

However, they had one unfortunate characteristic — a tendency to stray into the shipping lane. The shipping lane that passes south of Sri Lanka is one of the world's most important. It is a channel for all ships traveling from the Persian Gulf and the Red Sea (leading to the Suez Canal and Europe) to the Far East. The ships we sighted were almost all large and generally modern. There were high-sided auto carriers transporting Japanese cars to Europe and sickly green liquid natural gas transporters with large danger letters "L.N.G." on their sides, but most spectacular were the supertankers — enormous hulks with Arabian oil displacing volumes of water. These huge, almost uncontrollable ships are for me the most threatening inhabitants of the ocean. Plowing impersonally through the seas, unloaded offshore, they never touch land. We have seen dolphins leaping high from the pressure wave at their bows. The athletic dolphins, appearing tiny, are a striking contrast to the huge rigid forms that man has imposed upon the life of the oceans. One breach in the thin steel shell of a supertanker pours a suffocating layer of dark oil over hundreds of square miles of sea. Eight hundred feet behind the bow, cocooned by every luxury, the crew are insulated from the ocean on which they sail. Occasionally when they stop to unload, a dead whale is found wrapped around the bow; no one had noticed the impact. These behemoths could so easily crush *Tulip* and steam on unknowing. Lindy, not one for hiding her feelings, wrote in the log during her night watch: "Git these ships *outta* heah! They're terrorizing my watch."

The following night, she added: "Friggin' big fat tankers hoggin' up the ocean, causing me to lose the whales!"

Chris, characteristically, was more easygoing in his relationship with the big ships: "Yelled to the crew of a supertanker and invited them for breakfast. It was the least I could do after they stopped to avoid us!"

Among the sperm whales that we followed were some particularly distinctive individuals. One animal had large white patches on its back. A distant descendant of Moby-Dick, perhaps? There was also a small and inquisitive calf, whom Chris christened "Shred-head" because of some prominent scars on the right-hand side of its case. The gaps between the scratches suggested they had been made by the teeth of adult sperm whales. The scars were not deep and did not seem to be the results of aggression. The calf was probably "mouthed" as part of complex social interactions with the other members of its group.

Surprisingly, the sperm whale's narrow, fierce-looking jaw seems to be used primarily for social purposes. The squid that are found in the stomachs of freshly killed sperm whales rarely show teeth marks, and most appear to have been swallowed whole. There have also been a number of observations of healthy whales with broken or deformed jaws. Phil, while underwater, saw a whale whose lower jaw was broken about three-quarters of the way toward the tip. Another such animal was Moby-Dick, who scythed Ahab with his sickle-shaped jaw. Sperm whales have also been caught totally missing their lower jaws, but with stomachs full of squid.

If sperms do not bite their prey, how do they catch them? It seems likely they find the squid using echo-location, and suck them into their mouths with a powerful indraft of water. But this hydraulic suction cannot operate over more than a few yards at most. How does the sperm whale come within such close range of its victims? Does it simply chase them down? The old whalers had a theory that the vivid white lining of the sperm whale's mouth attracted the squid. Thomas Beale, the early-nineteenth-century ship's surgeon, described what was presumed to be the procedure:

> When the whale is inclined to feed, he descends a certain depth below the surface of the ocean, and there remains in as quiet a state as possible, opening his narrow elongated mouth until the lower jaw hangs down perpendicularly, or at right angles to the body. The roof of his mouth, the tongue, and especially the teeth, being of a bright glistening white colour, must of course present a remarkable appearance, which seems to be the incitement by which his prey are attracted.[1]

Our depth sounder records show sperm whales traveling

when at depth, and seeming to feed. This suggests that Beale's account is not always the whole truth. However, the extraordinarily bright white of the mouth of the sperm whale, which is so impressive when one is swimming with them, is likely to have some function. Like the lower jaw, whose principal purpose seems to be to touch, stroke, or in the case of the large males, battle other sperm whales, the white color of the mouth may have a social function. Animals often have bright colors on the parts of their bodies they use for signaling to one another. The bright red expandable throat of the magnificent frigate bird is a particularly splendid example. The white around the sperm whale's mouth, "glossy as bridal satins,"[2] may accentuate the importance of an open-mouthed display (of aggression, perhaps?) to another sperm whale.

Another theory on how the sperm whale catches its food has recently been proposed by Dr. Kenneth Norris of the University of California at Santa Cruz. He suggests that the whale may make particularly intense clicks in order to stun its prey.[3] We have listened for these "superclicks" through our hydrophones and once, last March, thought we heard them. However, Cedric Martenstyn, who was on board at the time, disillusioned us.

"Those loud bangs are probably fellows illegally dynamiting for tropical fish on the coral reefs off Trincomalee. They let off a charge, and the fish are stunned by the blast and float to the surface, where they are scooped up. The fish recover more or less, and are sold for the aquarium trade."

Well, at least Norris's basic theory does work in one instance! A colleague of mine, Dr. Peter Beamish, has pointed out that an overwhelming blast is not required. Certain sound frequencies can be extremely distracting to us, even though the sound may not be loud. Chalk scraping over a blackboard is one instance. Certain sounds may temporarily disorientate a squid long enough to allow a sperm whale to catch it.

A problem with this theory is that many squid seem to be deaf. When tested in tanks, they show no reaction to sounds of any type, including recordings of whale clicks. This seems strange: a species with even the slightest concept of natural selection would surely have evolved a way to sense the enormously loud, ill-boding clicks of such an important predator as the sperm whale. If a squid had even a very primitive "ear" so that it could sense the sperm whales at a few miles, as we can with our relatively insensitive hydrophones, and then took evasive action, surely it would greatly increase its "fitness" and be favored in natural selection. But squid do not appear to have read Darwin. An ingenious theory of why squid might be deaf has recently been put forward by Martin Moynihan. He suggests that squid may be deaf precisely to avoid the debilitating effects of Norris's "big bang" clicks.[4]

One of the principal objectives of this autumn season on board *Tulip* was to see if we could observe the large male sperm whales, who should have migrated from colder waters to mate, if those small calves we saw last March really were newborns, as the gestation period is about fifteen months. The observation of the birth of "Lindy's baby" and the other tiny calves in October suggested we had misplaced the whales' seasons and they were really mating in July, or perhaps there are two seasons or continuous breeding. However, we were still hoping to sight the large male sperms.

At midmorning on November 9, thirty miles southwest of Sri Lanka, Flip, whose sharp eyes put us all to shame, sighted some blows, and soon we came up to five whales lying at the surface.

"Oh, look! There's a big one in the middle!" cried Caroline.

"It's *enormous!*" added Phil. "Must be a full third longer than the others."

A fifty-foot-long male sperm whale lay in the center of the cluster, dwarfing the four females, two of which swam on either side of him. We just had the opportunity to marvel at his bulk and take a few photographs before the huge flukes were raised and he slipped beneath the waves. During the remaining thirty hours that we stayed with that particular group, we searched for the big male but never saw him again, despite several sightings of his erstwhile companions.

Two afternoons later, only twenty-five miles from Colombo, we were tracking some very loud slow clicks. They repeated every 5.5 seconds rather than every 0.6, which is the average rate for the females that we normally follow.

Suddenly Lindy cried, "Blow, and it's a big one! It must be a blue!"

"It's looks like it," I said. "But this is a strange place to find one. We're nowhere near the edge of the continental shelf."

We approached, but flukes were raised before we could come close enough to see the rest of the body clearly. However, when we listened on the hydrophone, clicks came blasting through. We started to suspect that this whale was not a blue. An hour and a quarter later, when it came to the surface again, we were right alongside a big male sperm. The whale was so much larger than the sperms we had become familiar with that he projected a completely different aura; he seemed to be of a different species. His head was swept with scars from battles with other males; his swimming was powerful and direct. At approximately forty-five tons, three times the mass of an adult female, the big bull surged through the waves. His slow click was lower and more intense than the clicks of the females. After the long, suspenseful pause, it drove through the hydrophones like a slammed jail-house door.

We followed "the Big Click" through the evening and into the night. At 9:00 p.m. the loud signature stopped. We waited. Just before midnight Phil picked up clicks to the north and Chris, who had the midnight to three "graveyard" watch, wrote in the log: "0100. Got in touch with El Toro, I think, and followed him into a group of good clickers. Bang, bang!"

What was going on down there? The big male seemed to have met up with a number of smaller sperm whales, probably a group of females. We heard many of the patterned codas that seem indicative of social interactions.

In the morning, when we sailed up to the whales, there were

females and a calf, but no sign of the big male. Like many another Lothario, "El Toro" had slipped away at dawn.

The two encounters with male sperm whales were frustrating because of their short duration, but even these fleeting glimpses were interesting. We now know that large male sperms do come to Sri Lankan waters in the autumn, but our observations suggest that there are not many of them and that they interact only briefly with groups of females. Could this be because of recent heavy whaling for large male sperms in antarctic waters, where "El Toro" presumably spends most of the year? Or do his fellows become sidetracked by other female groups farther south, never reaching the northern Indian Ocean?

The lives of large males are almost unstudied, but we can infer a little. They spend most of the year in cold waters, off Greenland or Iceland, in the northern North Pacific Ocean or near the antarctic ice. Males are often solitary, but sometimes they are seen in small groups. They feed much like the females whom we have studied from *Tulip*, but they probably dive deeper and attack larger squid or big fish. Do they use "the Big Click" during feeding, or is it reserved for their encounters with females or other males on the breeding grounds? We do not yet know, but this is an interesting question. Sometimes particular sounds, like the croak of the toad, can be reliable indicators of an animal's success in competition for mates. A deep croak signals a large and therefore formidable male toad. Perhaps "the Big Click" is also a signal of might. Females desiring mates or competing males might assess the potential of a particular male from some characteristic of his click. The click, whose original purpose was food finding, might then have evolved to emphasize these other characteristics.

Conventional theory has it that at the appropriate time every year, the mature males leave their feeding grounds and swim for warmer waters in search of females. But perhaps this is not so. Males have evolved to their extreme size pushed by sexual selection — a mating system in which the largest is the most successful. Unlike females, they are probably not optimally adapted for their ecological role; gathering enough food to maintain their huge bodies is no trivial task for fifty-five-foot males. This is probably why they spend most of the year in colder waters, where each piece of food, be it squid or fish, will be larger. A trip to the tropical mating grounds may be very costly energetically. Apart from the time and calories consumed in swimming to the tropics, in producing sperm, chasing females and possibly fighting other large males, the warmer waters, with their smaller squid, are less productive to the male than his usual polar haunts. Perhaps because of these energetic costs males might not breed every year. After completing a particularly taxing mating season, it might be a year or more before they have regained condition and are ready to tryst again. This is another possible cause for the infrequency with which we have sighted males.

Once on the mating grounds, what is the male's strategy? Does he take over a group of females and hold it for as long as he can, be it days or months? Or, as our two short observations suggest, does he interact only briefly with a group — just long enough to see if there are any available females — before moving on to another group?

Whalers often found more than one large male with a group of females. This has been interpreted in different ways. Were these males in the process of competing with one another, coincidentally "checking out" the same group of females, or was their relationship more long-term? The Soviet scientists D.D. Tormosov and E.G. Sazhinov have postulated that males may form coalitions on the breeding grounds — they might even be relatives from the same natal group — and enter the mating arena together.[5] Male lions have been found to use this type of cooperative mating strategy, which is successful because two males can better defend their sovereignty against intruders. The South African scientist Dr. Peter Best has suggested that a dominance hierarchy within male sperm whale coalitions, which will determine access to females, may be set up during aggressive encounters on the feeding grounds in advance of the breeding season.[6]

We need to know the details of the sperm whale mating system to have a realistic idea of their population dynamics and how the status of their stocks may be changing. If we can find an area with a reasonable number of male sperm whales interacting with groups of females, then I think that the techniques that have been developed on board *Tulip* will be the key.

On November 13 we put in to Colombo to keep an appointment with the film crew, who had made plans to return and take more footage of the sperms. Characteristically, there was no sign of them. It was a particularly trying port call. The rough seas in the Gulf of Mannar were entering Colombo Harbor, making our mooring far from comfortable. Trips to and from shore in the dinghy were equivalent to being dragged through a cesspool as the foul "water" sloshed over us. The engine alternator had broken, so we smuggled it through the customs post into town in order to get it rebuilt. Then there were problems with the finances for Jonathan's part of the study in the spring, so I had to make late-night phone calls to World Wildlife Fund in Switzerland.

N.A.R.A. also made difficulties for us: they had somehow become suspicious that we were going to sneak off with the film crew to film the whales without Sri Lankan input and had ordered the Sri Lankan Navy to apprehend us. Nothing was farther from my mind than to act illegally in this way, even if the film crew had been around. Luckily Sri Lanka does not boast one of the world's most effective navies. They failed to find us, even though we had innocently anchored one night outside one of their bases. I was shocked by N.A.R.A.'s total lack of trust, and I told them so. But they responded by demanding detailed itineraries and regular radio reports of our position. With our type of research and *Tulip*'s radio equipment, both demands were impossible to meet.

With no letup in the rough weather and our problems on-shore, we decided to sail toward the Maldive Islands, where the British Admiralty routing charts suggested we would find calmer weather. I could sympathize with the New Bedford whaling captain Samuel Braley of the ship *Harrison*, who, after three years on the Colombo grounds, in 1857 weighed anchor from the port of Colombo, pronounced it "one of the most miserable ones in the world"[7] and vowed never to return. I was disillusioned with Sri Lanka. My image of the country had shifted from the serene Buddhist monk to the confused and sad Tamil refugee.

For the first two days of our passage westward from Colombo we were too tired to consider trying to follow any whales. Reluctantly we passed over several groups of clicks. But with improving weather and the soothing lullaby of the ocean our energy returned.

Late on November 19 Lindy wrote in the log: "2100. Serene yet mystical moonlit bright night. Surreal illumination. Balmiest of balmy evenings.

"2300. Heard spermies! *Tulip* team spring into action. Hi, *babies*! Welcome back." (Note: Lindy uses the word *baby* to refer to both a young whale and to whales in general.)

For the next four days we sailed with groups of sperm whales in the area between the Maldive Islands and southwest India. The whales were generally easy to follow and we managed two fairly long trackings of particular groups. The weather was calm, with just an occasional gentle breeze. The northeast monsoon seemed to have been blocked by the high mountains of southern India. The sea was smooth and translucent. At night a full moon allowed us to watch the luminescent clouds of the whales' blows, as we heard the breathing of the great animals murmuring over the still seas. In this moon-drenched clarity we could even identify the lissome striped dolphins who one night came to ride *Tulip*'s bow. So calm was the sea that Chris remarked, "We've been had. It's just a big lake!"

Once we had left the rough seas of Sri Lanka and the fierce demands of land, the crew's spirits soared in these ideal conditions. One of the most enjoyable developments was Chris's invention of "Radio W.W.F." He had rigged up a communications system, originally designed by Jonathan, so that a crew member in the cockpit could, without shouting, communicate with a watcher up the mast. Lindy, who was at the spreaders scanning the horizon for blows, suddenly heard Chris's sugary disk-jockey accent coming through the headphones: "This is Radio W.W.F., the voice of *Tulip*, bringing you all the news and tunes from the Indian Ocean. Do *you* find it difficult to relate to inhabitants of the deep? Are you often left speechless when confronted by a wet dolphin snout? Well, stay tuned for our upcoming talk show 'Whales and You,' with experts from around the world. Our first distinguished guest is Dr. Linda Weilgart, who is speaking to us from halfway up *Tulip*'s mast.

"Dr. Weilgart, how can we establish a meaningful rapport with our marine cousins?"

"Well, Chris, that's a very interesting question. I believe that basically it is a matter of . . ."

"Absolutely, Dr. Weilgart. I think you have hit the nail on the head. And let *me* hit you with this popular tune from . . ."

Whereupon driving reggae would come blasting through Lindy's headphones as Chris pressed the communications system microphone to the earphone of his Walkman cassette tape recorder.

We were constantly watching and listening for large male sperm whales. One night Phil recorded an exceptionally loud series of clicks. As usual, he ended the tape recording with a voice-over:

"The time is 0100 on November 20, 1983. Our position is 6 degrees 48.3 minutes north, 75 degrees 32.2 minutes east, and if that wasn't 'the Big Click' I will eat this clipboard."

The following morning the other members of the crew took particular delight in timing the intervals between the clicks Phil had recorded — 0.9 seconds, close to the average interval for adult females, and certainly not a big male. Phil was offered various recipes for making clipboards more palatable.

Our time with the whales northeast of the Maldives was, as so often, limited by factors unrelated to the whales themselves. In this case our twin problems were lack of fuel and the burnout of our newly rebuilt alternator. The crew's notes in our log tell of a frustrating passage toward Male, the chief port and capital of the islands:

24 November 0900. Clarence is a worthless, unnecessary appendage. [Lindy] (Note: Clarence does not work with no wind.)
1000. Clarence has made a partial comeback. [Phil]
25 November 0230. Engine off. Ran out of diesel — oops. [Hal]
0300. Just as well. What kind of a sailboat *is* this? [Chris]
0400. Big jib up — Yeh! [Chris]
0500. Wind dying — ohh . . . [Chris]
0900. Zilch wind. [Chris]
1200. Becalmed in the drizzle. [Hal]
1300. No wind. *Desperate*. [Phil]
1400. The crew is becoming testy and unmanageable. Extreme action may need to be taken. The sharks await . . . [Lindy]
2000. *!!Z@| — Phil's comments. [Phil]
2100. Whattameal! Whatttawashup! [Chris]
26 November 0600. *Beautiful* sunrise and shooting stars. Gotta like it! [Lindy]
1600-1900. No entries in log.
1900. Crew suffers mental relapse and amnesiac stupor. [Lindy]
2000. The general consensus: Caroline's garlic bread was good. [Lindy]
2300. Incredible garlic bread Caroline. Oh, and no wind. [Phil]

27 November 0100. Inch by inch, yard by yard . . . [Chris, who has not fully absorbed the scientific imperative of the metric system.]
0200. Speed 2.0 knots. Knot bad. Knot what you'd expect, either. [Chris]
0600. A Maldivian fishin' vessel cruises by wanting no cigarettes and offering no diesel! [Caroline]
1200. Log imitating a dolphin off to port. [Lindy]
1700. Land ho! [Chris]

The Maldives is one of the smallest and strangest nations in the world. Physically it consists of thousands of tiny coral islands clustered into atolls. None of the islands is longer than two miles, and none rises more than several feet above the ocean. The islands are covered in palm trees and are fringed by white coral-sand beaches and reefs.

Although the Maldivian people, who number about one hundred and fifty thousand, probably originated from a racial stock similar to the Sinhalese, they are culturally very distinct. Entirely Muslim, and becoming more fundamentalist under the current government, Maldivians have their own language and even their own highly distinctive script, which is written on two levels from right to left. The people of these tiny islands depend on the sea. They fish and trade from their graceful "dhonis" — wide, flat, dhow-rigged boats, with proud, colorfully painted, upward-sweeping stems and sterns, much like Viking longships. Universally respected for their seamanship, Maldivians are valued as crew members on ships of many nations.

It took us a while to become used to the Maldivian way of going about things, so different was it from that of the Sri Lankans. On our first morning we sailed into the inner harbor of Male and anchored. This was absolutely forbidden, and unlike in Sri Lanka, where such rules can be overlooked if one makes a small bribe or argues vehemently enough, in the Maldives regulations are adhered to. The customs officer who confronted us was becoming almost apoplectic with rage. He seemed sincere in his threats to have us arrested, so we raised anchor and sailed out — a tricky procedure since the harbor was small, full of dhonis and other boats and *Tulip*'s batteries were so low we could not start the engine. It took a long time to find good holding ground for the anchor in the deep water of the outer harbor, but we finally discovered a place where *Tulip*'s anchor did not drag, sorted things out and rowed ashore for our first meal on dry land. We were confronted with another official, who told us we had to move *Tulip* again: she was in the line of sight of the Ministry of Defense Building, which was forbidden. This regulation appeared to be related to the Maldives being such a small country that a *Tulip*-size boat loaded with mercenaries could stage a meaningful coup. Once again we argued fruitlessly. Again arrest was threatened, so we rowed back out to *Tulip*, hauled the anchor aboard and searched for another spot where it might hold. When we finally ate our meal, we were all wondering if we had leaped from the Sri Lankan frying pan into the Maldivian fire.

However, once we had learned the rules, things proceded smoothly. Most of the government offices were in one building, and as long as we stayed within the regulations, the officials were courteous and efficient. The customs officer who had raged about our entry into the inner harbor became helpful and friendly. A local mechanic was able to sell us a spare alternator, which fit *Tulip*'s engine. The small scale of Male, the Maldivian capital — the whole island was less than a mile and a half long — meant that carrying out bureaucratic tasks and obtaining supplies took little time and produced no exasperation, a novel experience after Sri Lanka.

However, the Sri Lankans wanted us back. N.A.R.A. sent messages ordering us to return to Colombo to work with the film crew, who had finally arrived but could not find the whales without us. N.A.R.A. were loath to let the film be made outside Sri Lanka. I had no wish to sail back to Colombo: it would have left insufficient time to do any meaningful research and *Tulip*'s crew might have mutinied. The film crew's season was a complete waste and N.A.R.A. were offended by my stubbornness, but we were staying in the Maldives.

Although some Maldivians find the highly regulated Muslim society extremely restrictive — and I am sure we would, as well, if we lived there for any length of time — the honesty, efficiency, cleanliness and simplicity we encountered were most welcome.

The dhonis, almost the only means of transportation in the Maldives, sailed past our anchorage carrying fish, people, or whatever supplies were needed through the archipelago. As in Newfoundland, I rejoiced in being surrounded by a truly marine society. *Tulip*'s crew were able to snorkel among the Maldivian coral reefs. For the first time during this season, and for almost the first time during the whole *Tulip* Project, we could relax in port.

Recently the Maldivians have begun to exploit the simple beauty of their island home. The Maldivian government has set several islands aside as "tourist islands." Tourists can fly in from Europe to enjoy a simple holiday of sun, sea and diving. The Maldives has some of the best scuba diving in the world. By keeping tourists on prescribed islands, Maldivians minimize the impact on their traditional culture but gain the benefits of the foreign exchange that tourists bring.

Despite the tourist boom, fishing remains the way of life for many Maldivians. Dhonis put out into deep water off the islands and catch tuna by hook and line, using fish netted on coral reefs as bait. This form of fishing has few adverse side effects, as long as the bait fish on the local coral reefs are not overexploited. Most important, there are no dangerous artificial-fiber nets draped through the ocean, catching dolphins, turtles and anything else that comes their way. We were told that the Maldivian fishermen have few interactions with marine mammals. They used to hunt dolphins for shark bait, but that has now stopped.

In typically efficient Maldivian fashion, a statistical breakdown of the previous year's fishery was readily available (in contrast to Sri Lanka, where obtaining even a partial listing took hours of shuttling between government offices in Colombo). At the bottom of the list of fish species caught in the Maldives during 1982 was a curious entry: "Ambergris: 90 kg."

Ambergris is produced only by sperm whales. It seems to be formed by a cohesive lump of fecal material that builds up around undigestible squid beaks in the whale's intestines. Ambergris is usually employed as a fixative for perfumes, although, according to Frank Bullen, this is not its only use: "The Turks are said to use it for a truly Turkish purpose, which need not be explained here, while the Moors are credited with a taste for it in their cookery."[8] Because of these uses, ambergris is extremely valuable: whalers used to dream of making their fortunes by finding lumps of it in recently killed whales. However, this was rare; few sperm whale stomachs contain ambergris. Ambergris is occasionally vomited, or possibly defecated, and Maldivian fishermen keep a lookout for floating dark waxy lumps. If they find a lump of ambergris, it is the Maldivian equivalent of winning the lottery.

Ambergris from the Maldives is sold to Saudi Arabia. The use to which it is put is not specified in the Maldivian Fisheries Statistics. However, one aspect of this small part of the local economy that interested us was that ambergris was mostly found during the months of the northeast monsoon, from November to February, when it would probably be blown southwest toward the islands from the area in which we had been following sperm whales.

We are now heading back to the whales. Astern, a string of palm trees appears and then disappears as *Tulip* follows the contour of the waves. A land that can vanish because of less than three feet of vertical motion would seem ephemeral in the extreme. But, like many sights of land from sea, this is an illusion. There is a real country beneath those trees — the Republic of the Maldives.

13

TWO HUNDRED AND FIFTY MILES EAST OF THE MALDIVE ISLANDS

December 5 - 9

"HOW'S IT GOING, PHIL?" I ask, climbing out of the main hatch.

"There were some tense moments," he replies. "I nearly lost the whales twice, but the crises seem to have passed. The clicks are clear now."

We picked up a group of whales yesterday evening and have managed to track them through a rough and difficult night. They are the first sperm whales we have found since leaving Male five days ago. The wind has now moderated and Phil has *Tulip* motor-sailing after the clicks, with a gentle breeze on the beam filling the mainsail and working jib. But there is still an ugly sea left over from last night's blow, and the early morning sky is ominous. Imbedded in the light gray blanket of cloud are the dark bases of towering thunderheads. Distant thunder rumbles over the ocean, and a few miles to port, an indistinct waterspout joins the gray sea to the gray clouds.

"Are we close enough to see the sperms?" I ask Phil.

"No, not yet."

I go below to the chart table, where I take up a notebook in which I have been jotting down "useful things" for Jonathan: which parts of the engine need attention, weak points in the rigging, how to buy fuel in the Maldives, and so on. In another week our autumn season will be over, and we will leave *Tulip* in the Maldives under Phil's care until Jonathan takes over in late January.

"There's a squall coming," Phil calls out.

Immediately *Tulip* is thrown far over by the weight of the wind. As Phil struggles with the helm, I rush on deck and move forward to the mast in order to lower the mainsail. But there is a loud crack from above my head, then a drawn-out crash. The aluminum mast has snapped at its center and is tumbling over the side. The dismasting is complete in less than five seconds. The mast that has taken *Tulip* over so much ocean is a jumbled ruin hanging in the water to leeward. I am stunned by the suddenness of the transformation of the network of taut lines, wires and spars that framed our world to a complete mess.

"Is everyone okay? What happened?" calls out Lindy from down below as Phil looks down the hatch. "Are you all right down there?" Both are shocked, but their immediate concern is for the rest of the crew.

Caroline was struck on the back by a loose shroud that fell through the cabin skylight, but she does not seem to be badly injured. No one else was hurt.

The squall that brought the mast down soon passes. We gather on deck in the unreal calm.

"We'd better get that junk back on board."

If the seas increase, the flotsam could damage the hull. Phil puts on his snorkeling gear and dives over the side. He releases the sails from the tangled remains of the mast, and we haul them on board. Then comes the heavy work of heaving the two halves of the mast back on deck. Phil's strength has never been more valuable as we struggle with the torn aluminum. Without the stabilizing mast and sails, *Tulip* lurches awkwardly in the more valuable as we struggle with the torn aluminum. Without the stabilizing mast and sails, *Tulip* lurches awkwardly in the swell, and with no rigging to hold on to, moving around the deck is difficult.

It takes three hours to haul the carnage back on board, lash down the mast and boom, coil the rigging and stow the sails. I then assess our position. We are about two hundred fifty miles from Male, about the same distance from Sri Lanka and a little bit closer to India. Although *Tulip* has often sailed close to the Indian coast, we have so far avoided making port in India because of disquieting reports of the devious complexities of Indian bureaucracy. This is no time to investigate them. We choose Male as our destination. It is downwind and down-current, which will be important if our increasingly unreliable engine fails and we have to construct a jury-rig.

Tulip is a sorry sight — a base without a monument. Her decks are choked with twisted aluminum and coiled stainless-steel stays. Stanchions are broken, and in place of the tall, straight, smooth mast guiding the sweep of the sails, there stands a pathetic long-handled scrubbing brush, holding aloft the wind scoop, which brings ventilation down below, and a Maldivian "courtesy" flag.

I have left Jonathan a sad legacy: a broken mast, a crippled engine and an enraged N.A.R.A. I had so hoped he could begin his study with a clean slate, without the troubles that have plagued the start of each previous season. My little notebook of helpful hints is insignificant when compared to the greater disasters.

I look back at the two years since *Tulip* headed hopefully from the Suez Canal into the Indian Ocean. On one side are the fine times at sea: the exhausting but fascinating long watches of the families of sperm whales; the captivating grace of the whales when seen underwater; and hearing the laughter of tired friends while eating our evening meal in *Tulip*'s cockpit after a broiling day. These memories stemming from the sea hold me to this work and make my seasons ashore seethe with a longing to return.

But then there was the land. Having partially anticipated the bureaucracy, so grinding, petty and unnecessary, we generally managed to endure it. The manipulation of the *Tulip* Project by N.A.R.A. and other organizations who saw our research as a means of furthering their own, often political, goals was worse. We were like lake sailors who suddenly find themselves in the heart of a North Atlantic gale.

What does *Tulip* leave in her wake? Our principal objective was to find out whether we could study living sperm whales. Could work like that performed on *Tulip* substitute for data collected using the lethal techniques of the whalers? Could we penetrate the vital mysteries of the sperm whale, which the traditional methods based on whaling have no way of reaching?

Whaling is, thankfully, dying out. But one of the lingering arguments for its continuation is scientific: without whaling nothing would be known of whales. Roger Payne's research on right whales in Argentina, and the cooperative efforts of scientists studying humpback and gray whales around North America, have shown emphatically that for these three species at least, living whale science (benign research) is a match for the traditional techniques. But what about sperm whales? Our work on board *Tulip* has demonstrated that they, too, can be studied by benign research. Jonathan and I have looked at the different parameters of the sperm whale's existence needed to investigate their populations, such as the size of groups and the pregnancy rate of females. We predict that each of these parameters, which was formally estimated from catch statistics or examination of carcasses, can, with sufficient time and effort, come from the techniques developed on *Tulip*.

During many months of observation and experimentation, we have also found ways to examine some of the areas of sperm whale behavior where the whaling scientists are left guessing. We have followed groups of females for days, identified individuals within the groups and, on two occasions, watched or heard large males interact with groups of females. This is a beginning. When we can increase our time with a particular group from four days to more than ten, when we have repeatedly observed large males interacting with groups of females, perhaps even seen them mating, then we will begin to understand the system that ordains their relationships.

During these months at sea, I have watched the sperm whales, looking for keys to an understanding. I have found it impossible to function simply as an impassive machine turning the actions of the whales into scientific truths. Most of what I saw of the whales, what I felt when with them, will never reach the pages of learned journals — they remain within me. The sperm whale has taken his place in my mind as it did in Ahab's.

There are two images. The first is seen from above the water: a large dark cylinder, newly risen from a foreign domain, lies at the surface. The waves wash over the provocative form. What is its role in the depths where such bizarreness is almost customary? Why has it risen to jar our familiar world? Can it really be alive?

The second image is of the sperm whale underwater, just the head — the remainder of the body flows into the murk of the ocean beyond. There is the jaw, straight, firm and lined with white — closed now, the power concealed. Above stretches the forehead, its "lofty purpose" undisclosed, a symbol of all that the whale keeps hidden. Behind are the flippers, that steer, that touch. Closest is the eye of the whale. Timid and curious, it watches me.

Although *Tulip* looks a sorry shambles and this season's research is at an end, as we motor toward Male we stop each hour. I lower the hydrophone and hear the whales: "Click . . . click . . . click . . . ."

POSTSCRIPT

IN EARLY 1984 JONATHAN, GAY AND TWO new crew members, Nicola Davies and Vassili Papastavrou, flew out to join Phil, who had been looking after *Tulip* in the Maldives. They replaced the mast, smoothed over relations with N.A.R.A. and sailed back to Sri Lanka for the final season of the *Tulip* study. The letters I received from Trincomalee told of the joys of their work: long trackings of particular groups (the group containing "Scratch-face" was sighted in the same area we had seen them a year earlier) and swimming with whales. But they also had at least the usual dose of the frustrations of bureaucracy, publicity and an unreliable engine. In June 1984 Phil, Vassili and I ended the project by sailing *Tulip* back through the Red Sea and Suez Canal to France for a much-needed complete overhaul.

Jonathan, Gay and Roger Payne wanted to continue the Sri Lankan research, and planned to have Constellation Yachts build them a boat specially for sperm whale tracking. But the Sri Lankan civil war escalated, and with Trincomalee in the center of the violence, their plans have had to be postponed. Jonathan finished his Ph.D. in 1987, and has continued his sperm whale research by returning to the Azores, where he first studied them, on board the International Fund for Animal Welfare's boat *Song of the Whale*. As always, Jonathan has been energetically experimenting with new techniques, such as better directional hydrophones, photographic measurement methods and ways of collecting small pieces of skin from the whales. From this skin, DNA can be extracted, and from its analysis, we can learn of parentage, migrations and mating patterns.

I, too, wished to continue to sail with, and study, sperm whales, but I hoped to find a more convenient study area than Sri Lanka. Lindy and I (we were married in 1985) searched Townsend's charts of where the Yankee whalers made their kills, records of ocean weather conditions and government regulations. The Galápagos Islands off Ecuador had records of many sperm whales close to land, calm weather and well-defined protocols for visiting scientists. So *Tulip*, now back to her original name of *Elendil*, was sailed across the Atlantic and through the Panama Canal to the Galápagos, halfway around the world from Sri Lanka. During our first Galápagos study in 1985, we found circumstances almost ideal for our research: cool, calm waters with easily tracked groups of females as well as large males. With the help of other ex-*Tulip* crew members, Vassili Papastavrou, Margo Rice and Caroline Smythe, we have now made three studies off the Galápagos and collected large quantities of data on the sperm whales there. For instance, we now know over six hundred individuals from distinctive photographs of their flukes.

From these data we have been able to clarify and quantify many of the impressions we had during the *Tulip* voyages. Yes, large males do spend only a few hours with each group of females before moving on, and there are features of "the Big Click" that may help identify particular males. We have been able to look in some detail at the social relationships within the groups of females and calves, and to estimate population size and pregnancy rate. We now have the kinds of quantitative results that begin to convince International Whaling Commission scientists that it is worthwhile to study living sperm whales. But none of this would have been possible without the initial experiments on the *Tulip* voyage.

The world of the whales is rarely still. In the time that it has taken to write and revise this book, a virtual moratorium on whaling has been implemented by member countries of the International Whaling Commission. But Japan, Norway and Iceland, although having agreed to the moratorium, say they are now whaling "for scientific purposes," which is technically allowed under the International Whaling Commission charter. The need for scientific knowledge is their justification for economically lucrative slaughter. In 1989 Japan plans to take 875 whales "for science." The International Whaling Commission itself has condemned this "scientific whaling" and I hope that our work on sperm whales, which started on board *Tulip*, may help show that this is simply blatant prostitution of science. We can usually duplicate, and often improve upon, the scientific results obtained from carcasses while we sail with the whales and watch them live their lives. In addition to scientific data, and instead of the immediate returns of meat and money, our children, and their children, will also be able to sail and swim with whales, glimpsing the strange beings as they rise from the deep.

APPENDICES

Appendix I Glossary: Parts and Activities of Whales

Breach: A leap from the water showing forty to one hundred percent of the whale's body above the surface. The whale usually, but not always, reenters the water on its back.

Callus: A pale growth on the dorsal fin, indicative of a mature female sperm whale.

Case: The smooth "forehead" of the sperm whale, situated above the jaw and in front of the skull. It holds the "spermaceti organ," a large reservoir of oil.

Cetacean: A member of the order of marine mammals consisting of whales, dolphins and porpoises.

Cluster: As used in this book, members of a cluster of whales within one hundred yards of one another and with coordinating movements. Clusters have one to thirty members and a fluid composition.

Dorsal Fin: A fin situated on the back of the whale. On the sperm whale the dorsal fin is small.

Flippers: Two flat appendages on the lower forward part of the whale's body, one on either side. Flippers are anatomically analagous to the forelimbs on terrestrial mammals.

Fluke: The flat, horizontal tail of the whale.

Fluke-up: The action of raising the flukes into the air before a steep dive.

Group: As used in this book, sperm whales traveling together over periods of at least several days form a group. Groups of sperm whales usually contain ten to thirty animals. Generally synonymous with the whalers' terms "pod" or "school."

Lobtail: Thrash of the flukes downward onto the water surface.

Spyhop: A slow raising of part of the case out of the water, with the whale nearly vertical.

Appendix II   The *Tulip* Project: Principal Equipment

Tulip: 33-foot-long "Gladiateur" class sloop, built by Henri Wauquiez, France, in 1979

Engine: Volvo-Penta MD11C with "Sail-Drive"

Sails: Elvstrom, North Sails and Bruce Banks (who built the spinnaker "Wuff-Wuff")

Navigational Equipment: Tracor Transtar Satellite Navigator
Charles Hutchinson Sextant made in 1910
Simrad Skipper 603 Depth Sounder

Self-steering Gear ("Clarence"): Aries Vane Gear

Acoustic Equipment: Benthos AQ-17 Hydrophones
Ithaco 453 and Barcus-Berry Standard Preamplifiers
Uher 4200 Tape Recorders
Directional Hydrophones specially built by Memorial University and Dev-Tec Inc.
Tectronix Model R5110 Oscilloscope, used for "BBN-ing"

Photographic Equipment: Canon, Minolta, Nikon and Olympus 35 mm SLR Cameras; Lenses: 50 mm for mast-position measuring photographs, 80-200 mm zoom for dolphins, 300 mm for identification photographs; Autowinders and Motor Drives
Film: Mainly Ilford FP4 and Kodachrome 64
Nikonos II and III Underwater Cameras with 28 mm and 35 mm lenses

Video Equipment: Ferguson Videostar Color Camera and Recorder
Sony HVM Black-and-White Camera with Underwater Housing

Appendix III The Sperm Whale (*Physeter macrocephalus*)

Length and Weight: Adult males to 52 feet, 45 tons.
Adult females 33 feet, 15 tons.

Diving and Breathing: Sperm whales regularly dive to 1,300 feet and can reach at least 3,000 feet. They can remain underwater for over an hour. While at the surface the whales usually blow (breathe) at about 15-second intervals for approximately 10 minutes. The blow is forward and to the left.

Life Cycle: Sperm whales are born at a length of approximately 13 feet and a weight of 1 ton after a gestation period of about 15 months. Males reach sexual maturity at about 26 years, and females at about 9 years. Mature females give birth approximately every 5 years and suckle their offspring for 2 years or more.

Food: Mainly squid of various species (averaging about 3 pounds, but sometimes over 100 pounds), but also octopus and various fish.

Distribution: Over all deep ocean, but clustered near islands, shelf edges and other productive areas. Females with their young remain in or close to the tropical seas, but large males are found in polar waters.

World Population: Hundreds of thousands, considerably reduced by whaling.

Social Organization: Females travel in stable groupings of 8 to 35 animals together with their young. At about 7 years of age young males, and possibly females, as well, leave their natal groups to form bachelor groups. Larger males are generally found in smaller groupings and frequently alone.

Acoustic Output: Principally broad-band clicks (200-32,000 hertz), usually regularly spaced at about 0.6 second intervals, but sometimes grouped into "creaks" or "codas."

NOTES

Chapter 1

1. Herman Melville, *Moby-Dick* (Harmondsworth, U.K.: Penguin, 1972), 143.

2. Ibid., 279.

3. Ibid., 481.

4. Ibid., 486.

5. Ibid., 487.

Chapter 2

1. Thomas Beale, *The Natural History of the Sperm Whale* [1839] (London: Holland Press, 1973), 191.

2. Melville, *Moby-Dick*, 481-482.

3. Mathew F. Maury, *A Chart Showing the Favourite Resort of the Sperm and Right Whale* (U.S. Navy: 1852).

4. Charles H. Townsend, "The distribution of certain whales as shown by the logbook records of American whaleships," in *Zoologica*, vol. 19 (New York: 1935), 1-50.

5. W.F. Mörzer Bruyns, *Field Guide to the Whales and Dolphins* (Amsterdam: Uitgeverij tor/n v. Uitgeverij V.H.C.A. mees Zeiseniskade, 1971).

6. Melville, *Moby-Dick*, 98.

7. Ibid., 381.

Chapter 3

1. Hydrographer of the Navy, *Red Sea and Gulf of Aden Pilot*, 12th ed. (London: 1980), 11.

2. Melville, *Moby-Dick*, 486.

3. Ibid., 301.

4. Ibid., 235.

5. Frank T. Bullen, *Creatures of the Sea* (London: The Religious Tract Society, 1904), 56.

6. Roger S. Payne and Scott McVay, "Songs of Humpback Whales," in *Science*, vol. 173 (Washington: 1971), 583-597.

7. Peter Tyack, "Interactions between singing humpback whales and conspecifics nearby," in *Behavioural Ecology and Sociobiology*, vol. 8 (Berlin: 1981), 105-116.

8. Roger Payne and Linda N. Guinee, "Humpback whale songs as an indicator of 'stocks,'" in *Communication and Behavior of Whales*, ed. Roger S. Payne (Boulder, Colorado: Westview Press, 1983), 333-358.

9. Katherine Payne, Peter Tyack and Roger Payne, "Progressive changes in the songs of humpback whales (*Megaptera novaeangliae*); a detailed analysis of two seasons in Hawaii," in *Communication and Behavior of Whales*, ed. Roger S. Payne (Boulder, Colorado: Westview Press, 1983), 9-57.

Chapter 4

1. Melville, *Moby-Dick*, 388.

2. Ronald W. Keller, Stephen Leatherwood and Sidney J. Holt, "Indian Ocean Cetacean Survey, Seychelle Islands, April through June 1980," in *Reports of the International Whaling Commision*, vol. 32 (Cambridge, U.K.: 1982), 503-513.

3. Seiji Ohsumi, "Some investigations of the school structure of the sperm whale," in *Scientific Reports of the Whales Research Institute of Tokyo*, vol. 23 (Tokyo: 1971), 1-25.

4. Frank T. Bullen, *The Cruise of the Cachalot* (New York: Dodd, Mead, 1926), 86.

5. Melville, *Moby-Dick*, 443.

6. Kenneth S. Norris and George W. Harvey, "A theory for the function of the spermaceti organ of the sperm whale (*Physeter catodon* L.)," in *Animal Orientation and Navigation*, ed. S.R. Galler et al. (Washington: NASA, 1972), 397-417.

7. Malcolm R. Clarke, "Function of the spermaceti organ of the sperm whale," in *Nature*, vol. 228 (London: 1970), 873-874.

8. Beale, *The Natural History of the Sperm Whale*, 48.

9. Melville, *Moby-Dick*, 667.

Chapter 5

1. E.J. Linehan, "The trouble with dolphins," in *National Geographic*, vol. 155 (Washington: 1979), 506-540.

Chapter 6

1. Melville, *Moby-Dick*, 298.

Chapter 7

1. Ø. Olsen, "On the external character and biology of Bryde's whale (*Balaenoptera brydei*), a new rorqual from the coast of South Africa," in *Proceedings of the Zoological Society of London*, (London: 1913), 1073-1090.

2. Melville, *Moby-Dick*, 236.

3. Toshio Kasuya and Seiji Ohsumi, "A secondary sexual character of the sperm whale," in *Scientific Reports of the Whales Research Institute of Tokyo*, vol. 20 (Tokyo: 1966), 89-94.

4. Melville, *Moby-Dick*, 496-497.

5. William A. Watkins and William E. Schevill, "Sperm whale codas," in *Journal of the Acoustical Society of America*, vol. 62 (New York: 1977), 1475-1490.

Chapter 8

1. Beale, *The Natural History of the Sperm Whale*, 5-6.

2. Melville, *Moby-Dick*, 307.

3. Peter Tyack and Hal Whitehead, "Male competition in large groups of wintering humpback whales," in *Behaviour*, vol. 83 (Leiden: 1983), 132-154.

Chapter 9

1. Robert L. Trivers, "The evolution of reciprocal altruism," in *Quarterly Review of Biology*, vol. 46 (Baltimore: 1971), 35-57.

2. Masaharu Nishiwaki, "Aerial photographs showing sperm whales' interesting habits," in *Norsk Hvalfangsttidende*, vol. 51 (Oslo: 1962), 393-398.

3. Bullen, *The Cruise of the Cachalot*, 212.

Chapter 11

1. R.W. Sheldon, A. Prakash and W.H. Sutcliffe, Jr., "The size distribution of particles in the ocean," in *Limnology and Oceanography*, vol. 17 (Lawrence, U.S.: 1972), 327-340.

2. John W. Kanwisher and Sam H. Ridgway, "The physiological ecology of whales and porpoises," in *Scientific American*, vol. 248 (New York: 1983), 110-121.

3. Melville, *Moby-Dick*, 371.

4. Ibid., 454.

5. Ibid., 497.

6. Ray Gambell, Christina Lockyer and Graham J.B. Ross, "Observations on the birth of a sperm whale calf," in *South African Journal of Science*, vol. 69 (Johannesburg: 1973), 147-148.

Chapter 12

1. Beale, *The Natural History of the Sperm Whale*, 35.

2. Melville, *Moby-Dick*, 431.

3. Kenneth S. Norris and Bertel Mohl, "Can odontocetes debilitate prey with sound?" in *American Naturalist*, vol. 122 (Chicago: 1983), 85-104.

4. Martin Moynihan, "Why are cephalopods deaf?" in *American Naturalist*, vol. 125 (Chicago: 1985), 465-469.

5. D.D. Tormosov and E.G. Sazhinov, "Nuptial behaviour in *Physeter catodon*," in *Zoologicheskii Zhurnal*, vol. 53 (Moscow: 1974), 1105-1106.

6. Peter B. Best, "Social organization in sperm whales, *Physeter macrocephalus*," in *Behavior of marine animals*, vol. 3., ed. H.E. Winn and B.L. Olla, (New York: Plenum Press, 1979), 227-290.

7. Phoebe Wray and Kenneth R. Martin, "Historical whaling records from the western Indian Ocean," *International Whaling Commission Scientific Committee Document No. SC/32/O8* (Cambridge, U.K.: 1980).

8. Beale, *The Cruise of the Cachalot*, 63.

SUGGESTED READING

General Books about Whales (including Field Guides)

Burton, R. 1980. *The life and death of whales*. New York: Universe Books.

Ellis, R. 1980. *The book of whales*. New York: Knopf.

Evans, P.G.H. 1987. *The natural history of whales and dolphins*. New York: Facts on File.

Gaskin, D.E. 1982. *The ecology of whales and dolphins*. London: Heinemann.

Hoyt, E. 1984. *The whale watcher's handbook*. Markham, Canada: Penguin.

Katona, S., V. Rough and D. Richardson. 1983. *A field guide to the whales, porpoises and seals of the Gulf of Maine and eastern Canada*. New York: Scribner's.

Leatherwood, S., R.R. Reeves and L. Foster. 1983. *The Sierra Club handbook of whales and dolphins*. San Francisco: Sierra Club Books.

Leatherwood, S., R.R. Reeves, W.F. Perrin and W.E. Evans. 1982. *Whales, dolphins and porpoises of the eastern North Pacific and adjacent Arctic waters. A guide to their identification*. NOAA Technical Report. NMFS Circular 444.

Mackintosh, N. 1965. *The stocks of whales*. London: Fishing News Books.

Matthews, L.H. 1978. *The natural history of the whale*. New York: Columbia University Press.

McIntyre, J. 1974. *Mind in the waters*. New York: Scribner's.

Minasian, S.M., K.C. Balcomb III and L. Foster. 1984. *The world's whales. The complete illustrated guide*. Washington: Smithsonian Books.

Mörzer Bruyns, W.F. 1971. *Field guide to the whales and dolphins*. Amsterdam: Uitgeverij tor/n v. Uitgeverij V.H.C.A. mees Zeiseniskade.

Ommaney, F.D. 1971. *Lost leviathan, whales and whaling*. New York: Dodd, Mead.

Slijper, E.J. 1958. *Whales*. London: Hutchinson.

Watson, L. 1981. *Sea guide to whales of the world*. London: Hutchinson.

Collections of Scientific Papers on Whales

Andersen, H.T., ed. 1969. *The biology of marine mammals*. New York: Academic Press.

Food and Agriculture Organization of the United Nations (FAO). 1981. *Mammals of the seas*. Vols. 1-3. Rome: FAO.

Herman, L.M., ed. 1980. *Cetacean behavior*. New York: Wiley.

International Whaling Commission. *Scientific Reports and Special Issues*. Cambridge, UK: IWC.

Norris, K.S., ed. 1966. *Whales, dolphins and porpoises*. Berkeley, Calif.: Univ. of Calif. Press.

Payne, R.S., ed. 1983. *Communication and behavior of whales*. Boulder, Colorado: Westview Press.

Ridgway, S.H., ed. 1972. *Mammals of the sea*. Springfield, Illinois: Charles C. Thomas.

Schevill, W.E., ed. 1974. *The whale problem*. Cambridge, Mass: Harvard Univ. Press.

Winn, H.E. and B.L. Olla., eds. 1979. *Behavior of marine animals*. Vol. 3, *Cetaceans*. New York: Plenum Press.

Historical Books on Sperm Whales

Beale, T. [1839] 1973. *The natural history of the sperm whale*. London: Holland Press.

Bennett, F.D. 1840. *Narrative of a whaling voyage around the globe from the year 1833 to 1836*. London: Richard Bentley.

Bullen, F. 1899. *The cruise of the Cachalot*. London: Smith, Elder.

Melville, H. [1851] 1964. *Moby-Dick or the white whale*. Indianapolis: Bobbs-Merrill.

Scammon, C.M. 1874. *Marine mammals of the northwestern coast of North America*. San Francisco: John H. Carmany.

Scoresby, W. 1820. *An account of the Arctic regions with history and a description of the northern whale fishery*. Edinburgh: Archibald Constable.

A Few of the More Important Scientific Papers on Sperm Whales

Backus, R.H. and W.E. Schevill. 1966. *Physeter* clicks. In *Whales, dolphins and porpoises*, ed. K.S. Norris, 510-527. Berkeley: Univ. of Calif. Press.

Best, P. 1979. Social organization in sperm whales, *Physeter macrocephalus*. In *Behavior of marine animals*, Vol. 3, ed. H.E. Winn and B.L. Olla, 227-290. New York: Plenum Press.

Best, P.B., P.A.S. Canham and N. Macleod. 1984. Patterns of reproduction in sperm whales, *Physeter macrocephalus*. *Reports of the International Whaling Commission* (Special Issue 6):51-79.

Caldwell, D.K., M.C. Caldwell and D.W. Rice. 1966. Behavior of the sperm whale. In *Whales, dolphins and porpoises*, ed. K.S. Norris, 677-717. Berkeley: Univ. of Calif. Press.

Clarke, M.R. 1970. Function of the spermaceti organ of the sperm whale. *Nature* 228:873-874.

Clarke, M.R. 1980. Cephalapoda in the diet of sperm whales of the southern hemisphere and their bearing on sperm whale biology. *Discovery Reports* 37:1-324.

Matthews, L.H. 1938. The sperm whale, *Physeter catodon. Discovery Reports* 17:95-164.

Norris, K.S. and G.W. Harvey. 1972. A theory for the function of the spermaceti organ of the sperm whale (*Physeter catodon* L.). In *Animal orientation and navigation*, ed. S.R. Galler, K. Schmidt-Koenig, G.J. Jacobs and R.E. Belleville, 397-417. Washington: NASA.

Townsend, C.H. 1935. The distribution of certain whales as shown by the logbook records of American whaleships. *Zoologica* 19:1-50.

Watkins, W.A. and W.E. Schevill. 1977. Sperm whale codas. *Journal of the Acoustical Society of America* 62:1475-1490.

Magazine Articles and Scientific Papers on the *Tulip* Project and the Living Sperm Whale Research that Followed It

Alling, A. 1985. Remoras and blue whales: A commensal or mutual interaction? *Whalewatcher* (Journal of the American Cetacean Society) 19:16-19.

Alling, A. 1986. Records of odontocetes in the northern Indian Ocean (1981-1982) and off the coast of Sri Lanka (1982-1984). *Journal of the Bombay Natural History Society* 83:376-394.

Arnbom, T. 1987. Individual identification of sperm whales. *Reports of the International Whaling Commission* 38:201-204.

Arnbom, T., V. Papastavrou, L.S. Weilgart and H. Whitehead. 1987. Sperm whales react to an attack by killer whales. *Journal of Mammalogy* 68:450-453.

Arnbom, T. and H. Whitehead. 1989. Observations on the composition and behaviour of groups of female sperm whales near the Galápagos Islands. *Canadian Journal of Zoology* 67:1-7.

Gordon, J.C.D. 1987. Behaviour and ecology of sperm whales off Sri Lanka. Ph.D. thesis, University of Cambridge, Cambridge, England.

Gordon, J.C.D. 1987. Sperm whale groups and social behaviour observed off Sri Lanka. *Reports of the International Whaling Commission* 37:205-217.

Lagendijk, P. 1982. Blast! *Panda* (W.W.F. The Netherlands).

Mullins, J., H. Whitehead and L.S. Weilgart. 1988. Behaviour and vocalizations of two single sperm whales, *Physeter macrocephalus*, off Nova Scotia. *Canadian Journal of Fisheries and Aquatic Sciences* 45:1736-1743.

Papastavrou, V., S.C. Smith and H. Whitehead. 1989. Diving behaviour of the sperm whale, *Physeter macrocephalus*, off the Galápagos Islands. *Canadian Journal of Zoology* 67:839-846.

Rice, M. 1984. The *Tulip* in the Indian Ocean Marine Sanctuary: the second season. *Whalewatcher* (Journal of the American Cetacean Society) 18:14-16.

Weilgart, L.S. and H. Whitehead. 1986. Observations of a sperm whale (*Physeter catodon*) birth. *Journal of Mammalogy* 67:399-401.

Weilgart, L.S. and H. Whitehead. 1988. Distinctive vocalizations from mature male sperm whales (*Physeter macrocephalus*). *Canadian Journal of Zoology* 66:1931-7.

Whitehead, H. 1982. Sperm whales off Sri Lanka. *Loris* (Journal of the Wildlife and Nature Protection Society of Sri Lanka) 16:23-24.

Whitehead, H. 1983. Baleen whales off Sri Lanka. *Loris* (Journal of the Wildlife and Nature Protection Society of Sri Lanka) 16:176-181.

Whitehead, H. 1984. Sperm whale — the unknown giant. *National Geographic* 166:774-789.

Whitehead, H. 1985. Benign research on sperm whales in the Indian Ocean Sanctuary. *Whalewatcher* (Journal of the American Cetacean Society) 19:3-9.

Whitehead, H. 1985. Humpback whale songs from the northern Indian Ocean. *Investigations on Cetacea* (Bern, Switzerland) 17:157-162.

Whitehead, H. 1986. Call me gentle. *Natural History* 95:4-11.

Whitehead, H. 1987. Social organization of sperm whales off the Galápagos: Implications for management and conservation. *Reports of the International Whaling Commission* 37:195-199.

Whitehead, H. 1987. Sperm whale behavior on the Galápagos Grounds. *Oceanus* 30:49-53.

Whitehead, H. 1989. Foraging formations of Galápagos sperm whales. *Canadian Journal of Zoology* 67.

Whitehead, H. and T. Arnbom. 1987. Social organization of sperm whales off the Galápagos Islands, February-April 1985. *Canadian Journal of Zoology* 65:913-919.

Whitehead, H. and J. Gordon. 1986. Methods of obtaining data for assessing and modelling sperm whale populations which do not depend on catches. *Reports of the International Whaling Commission* (Special Issue 8):149-166.

Whitehead, H., V. Papastavrou and S.C. Smith. 1989. Feeding success of sperm whales and sea-surface temperatures off the Galápagos Islands. Marine Ecology Progress Series.